CLEANING UP SITES CONTAMINATED WITH RADIOACTIVE MATERIALS

INTERNATIONAL WORK[

Glenn E. Schweitzer, Frank L. Parker, and Kelly Robbins, *Editors*

Committee on Cleaning Up of Radioactive Contamination:
Russian Challenges and U.S. Experience

Office for Central Europe and Eurasia
Development, Security, and Cooperation
Policy and Global Affairs

NATIONAL RESEARCH COUNCIL
OF THE NATIONAL ACADEMIES

In cooperation with the Russian Academy of Sciences

THE NATIONAL ACADEMIES PRESS
Washington, D.C.
www.nap.edu

THE NATIONAL ACADEMIES PRESS 500 Fifth Street, N.W. Washington, DC 20001

NOTICE: The project that is the subject of this report was approved by the Governing Board of the National Research Council, whose members are drawn from the councils of the National Academy of Sciences, the National Academy of Engineering, and the Institute of Medicine. The members of the committee responsible for the report were chosen for their special competences and with regard for appropriate balance.

This study was supported by a grant from the Russell Family Foundation to the National Academy of Sciences. Any opinions, findings, conclusions, or recommendations expressed in this publication are those of the author(s) and do not necessarily reflect the views of the organizations or agencies that provided support for the project.

International Standard Book Number-13: 978-0-309-12761-5
International Standard Book Number-10: 0-309-12761-0

A limited number of copies are available from the Office for Central Europe and Eurasia, National Research Council, 500 Fifth Street, N.W., Washington, DC 20001; (202) 334-2376.

Additional copies of this report are available from the National Academies Press, 500 Fifth Street, N.W., Lockbox 285, Washington, DC 20055; (800) 624-6242 or (202) 334-3313 (in the Washington metropolitan area); Internet, *http://www.nap.edu*

Printed in the United States of America.

THE NATIONAL ACADEMIES

Advisers to the Nation on Science, Engineering, and Medicine

The **National Academy of Sciences** is a private, nonprofit, self-perpetuating society of distinguished scholars engaged in scientific and engineering research, dedicated to the furtherance of science and technology and to their use for the general welfare. Upon the authority of the charter granted to it by the Congress in 1863, the Academy has a mandate that requires it to advise the federal government on scientific and technical matters. Dr. Ralph J. Cicerone is president of the National Academy of Sciences.

The **National Academy of Engineering** was established in 1964, under the charter of the National Academy of Sciences, as a parallel organization of outstanding engineers. It is autonomous in its administration and in the selection of its members, sharing with the National Academy of Sciences the responsibility for advising the federal government. The National Academy of Engineering also sponsors engineering programs aimed at meeting national needs, encourages education and research, and recognizes the superior achievements of engineers. Dr. Charles M. Vest is president of the National Academy of Engineering.

The **Institute of Medicine** was established in 1970 by the National Academy of Sciences to secure the services of eminent members of appropriate professions in the examination of policy matters pertaining to the health of the public. The Institute acts under the responsibility given to the National Academy of Sciences by its congressional charter to be an adviser to the federal government and, upon its own initiative, to identify issues of medical care, research, and education. Dr. Harvey V. Fineberg is president of the Institute of Medicine.

The **National Research Council** was organized by the National Academy of Sciences in 1916 to associate the broad community of science and technology with the Academy's purposes of furthering knowledge and advising the federal government. Functioning in accordance with general policies determined by the Academy, the Council has become the principal operating agency of both the National Academy of Sciences and the National Academy of Engineering in providing services to the government, the public, and the scientific and engineering communities. The Council is administered jointly by both Academies and the Institute of Medicine. Dr. Ralph J. Cicerone and Dr. Charles M. Vest are chair and vice chair, respectively, of the National Research Council.

www.national-academies.org

NATIONAL RESEARCH COUNCIL
COMMITTEE ON CLEANING UP OF RADIOACTIVE CONTAMINATION: RUSSIAN CHALLENGES AND U.S. EXPERIENCE

Frank L. Parker, *Chair*, Vanderbilt University
Brian B. Looney, Savannah River National Laboratory
Dieter K. Rudolph, Science Applications International Corporation

Staff

Glenn E. Schweitzer, Program Director, National Research Council
A. Chelsea Sharber, Senior Program Associate, National Research Council
Kelly Robbins, Senior Program Officer, National Research Council

Contents

Other Contributions

Appendixes

Preface

The National Academies and the Russian Academy of Sciences, with the support of the Russell Family Foundation and the International Science and Technology Center, organized an international workshop in Moscow on June 4-6, 2007, on *Cleaning Up Sites Contaminated with Radioactive Materials*. The emphasis was on sites in Russia, although experiences in cleaning up sites in the United States, Belarus, and Kazakhstan added an important dimension to the discussions. The purposes of the workshop were to exchange information on approaches and problems in addressing contaminated sites, to review recent progress and next steps in cleaning up some of the worst polluted sites in Russia, and to stimulate greater attention within Russia to the severity of the problems and the urgency in cleaning up a number of sites that are of concern to nearby populations and to the international community.

More than 100 officials and specialists participated in the workshop. The Russian participants were from important organizations that have been playing key roles in site remediation activities. The specialists from the United States, Belarus, and Kazakhstan also have had extensive experience in addressing closely related problems in their countries.

The initial sessions of the workshop were devoted to general overviews of site remediation challenges in Russia and the United States. Subsequent Russian presentations were directed to case studies of contaminated sites in Russia.

American specialists responded to each of these presentations from their perspectives and experiences in addressing similar problems. Other presentations highlighted persistent problems in the former Soviet Union (e.g., the aftermath of the Chernobyl accident) and opportunities offered by new technologies that can assist in cleanup activities (e.g., geographic information systems). Collectively, these presentations provide a strong basis for urging Russian organizations to devote greater resources to cleanup activities in the near term.

In addition to the formal presentations set forth in this report, additional papers and extended abstracts on directly related topics were prepared by the participants and made available during the workshop. Identified in Appendix B, they may be obtained by contacting the authors directly at the indicated e-mail addresses.

ACKNOWLEDGMENTS

The workshop and this publication were supported by a grant from the Russell Family Foundation. Since 2002, the foundation has been a strong supporter of the joint activities of the National Academies and the Russian Academy of Sciences directed to reducing problems associated with radioactive waste. The International Science and Technology Center, which has supported many cleanup activities in the former Soviet Union for more than a decade, provided supplemental funds for the workshop.

The statements made in the enclosed papers are those of the individual authors and do not necessarily represent positions of the Russell Family Foundation, the National Academies, the Russian Academy of Sciences, or other organizations where the authors are employed.

This volume has been reviewed in draft form by individuals chosen for their technical expertise, in accordance with procedures approved by the National Research Council's Report Review Committee. The purpose of this independent review is to provide candid and critical comments that will assist the institution in making its published report as sound as possible and to ensure that the report meets institutional standards for quality. The review comments and draft manuscript remain confidential to protect the integrity of the process.

We wish to thank the following individuals for their review of selected papers: Ann Clarke, ANC Associates; Keith Compton, Nuclear Regulatory Commission; Ken Czerwinski, University of Nevada, Las Vegas; Rodney Ewing, University of Michigan; Loren Habegger, Argonne National Laboratory; Milton Levenson, Bechtel International (Retired); and Bruce Napier, Pacific North West National Laboratory.

Although the reviewers listed above have provided constructive comments and suggestions, they were not asked to endorse the content of the individual papers. Responsibility for the final content of the papers rests with the individual authors.

Special thanks are extended to Kelly Robbins for her translation of some of the Russian language papers into English and to Jan Dee Summers for editing the proceedings.

Frank L. Parker, *Chair*
National Research Council Committee on Cleaning Up of Radioactive Contamination: Russian Challenges and U.S. Experience

Glenn E. Schweitzer, *Director*
Office for Central Europe and Eurasia, National Research Council

Opening Remarks

1

Welcoming Remarks

Nikolay Laverov, Russian Academy of Sciences

In Russia the most serious contamination of territories with radioactive materials is due primarily to development and fabrication of nuclear weapons, submarines, and icebreakers. First and foremost, we are concerned about the areas hosting (1) the Mayak reprocessing facility, (2) navy bases in the north European part of Russia and the Pacific Coast region, and (3) Kara Sea hot spots adjacent to Novaya Zemlya, where damaged nuclear submarines, the reactor of the icebreaker *Lenin*, and compacted radioactive waste have been deposited.

In past years, research has been carried out to facilitate efforts to clean up contaminated territories, develop new decontamination technologies, and reduce radionuclide levels in reservoirs like Karachai Lake, the Techa Cascade, the Kola fjords, and elsewhere. Meanwhile, the Principles of State Policy for Elimination of Radiation Hazards that were approved by President Vladimir Putin have defined objectives aimed at removing the waste that accumulated during the arms race in the Soviet era and that resulted from peaceful uses of atomic energy. A long-term program (up to 2020) on radioactive waste management and rehabilitation of contaminated areas has been developed.

I hope that at this workshop participants will present new findings on improved technologies for such work and that new opportunities for joint collaboration will emerge to promote the application of knowledge, expertise, and practical experience existing not only in Russia but in the United States and other countries as well. I wish all participants every success in carrying out their seminal work at the workshop.

2

Welcoming Remarks

Frank L. Parker, Vanderbilt University

On behalf of the U.S. National Academies, I am pleased to welcome you to this International Workshop on Cleaning Up Sites Contaminated with Radioactive Materials, organized by the U.S. National Academies and the Russian Academy of Sciences. This workshop is a continuation of a series of joint workshops carried out by the academies in the field of nuclear science and technology.

Recent interacademy activities on disposal of radioactive material have included two earlier workshops devoted to the proposed International Site for Storage of Spent Nuclear Fuel. The analyses of the location and operation of this potential site in Siberia have provided important information that is relevant to widespread interest in consolidation of spent nuclear fuel on an international basis. Those workshops were also supported by the Russell Family Foundation. In addition, joint studies supported by other sponsors have been directed to (1) U.S.-Russian collaboration in combating radiological terrorism, which was completed in 2007 and emphasized appropriate stewardship, including disposal of ionizing radiation sources; and (2) many aspects of the nuclear fuel cycle, a study that is still under way. Indeed, since the late 1990s, more than 10 interacademy workshops and joint studies have addressed various aspects of developments in the nuclear field.

It is important for the governments of the countries represented at this workshop to have independent scientific and engineering advice from their top experts on issues of economic and security importance. My colleagues from the United States and I are delighted to be here. We look forward to our interactions with our colleagues from Russia, Belarus, and Kazakhstan and from the International Science and Technology Center as we strive to increase mutual understanding and collaboration in a critically important field.

3

Welcoming Remarks

David N. McNelis, University of North Carolina

On behalf of George Russell, the Russell Family Foundation, and myself, I wish also to extend a welcome to this workshop. Since 1992, George Russell has promoted cooperation that will help develop mutually beneficial economic and political relations between the United States and Russia. Of special interest is the large amount of highly radioactive materials requiring final disposition that have accumulated in both countries.

Our primary focus is on the transmutation of components of spent nuclear fuel. But our interests more broadly have included approaches to nonproliferation of nuclear material, repository thermal loading, multinational approaches to fuel cycle management, development of risk assessment and risk management techniques, and with this workshop, remediation of sites with radioactive contamination.

As a sponsor of this workshop, we look forward to the proceedings of the meeting, with site-specific recommendations from individual participants on remediation activities and technologies. The proceedings should be brought to the attention of the officials responsible for advising on and establishing national priorities and budgets. I hope that the effort will result in appropriate priority being given to the remediation of each of the sites that have been identified as case studies.

I wish you every success in your discussions.

4

Interests of the International Science and Technology Center

Norbert Jousten, International Science and Technology Center

I am glad to welcome participants to this international workshop. It is a pleasure to see such representative groups of scientists from the United States and Russia, and also specialists from Belarus.

The International Science and Technology Center (ISTC) was created in 1992 to

- provide former weapons scientists in Russia and the Commonwealth of Independent States (CIS) countries the opportunity to redirect their knowledge and skills to peaceful activities;
- support basic and applied research and technology development;
- contribute to the transition to market-based economies;
- foster the integration of former weapons scientists and engineers from Russia and CIS countries into the global scientific community; and
- contribute to solving national and international technical problems.

The geographical footprint of ISTC membership has expanded greatly. While the original members were limited to the Russian Federation, the European Union, Japan, and the United States, today nearly 40 countries participate in ISTC activities. The ISTC coordinates the efforts of numerous governments, international organizations, and private sector industry, providing former weapons scientists from Russia and the CIS new opportunities for international partnerships. Through its political, legal, and financial frameworks, the ISTC contributes to fundamental research, international nonproliferation programs, and innovation and commercialization by linking the demands of international markets with the exceptional pool of scientific talent available in Russian and CIS institutes.

The topic of this workshop falls within one of the most important technology areas of interest to the ISTC, the *environment*. About 16 percent of ISTC funding of more than $750 million has been devoted to this area, with a number of projects of direct relevance to this workshop.

The most important results of these relevant projects relate to creation of radiological databases, as they are foundations for rehabilitation of sites contaminated with radioactive materials. The following databases were created: conditions in the territory of the former USSR (in the frameworks of ISTC projects 245, 245-2B, 245-2C, 2097), in the seas surrounding the territory of Russia (projects 101, 101-2, 101-3), in the Tobol and Irtysh rivers (near the Mayak Production Association, project 2558), and at the Semipalatinsk nuclear test site in Kazakhstan (projects K-414, K-414-2, K-1125).

Let me wish the participants success in this workshop that is important not only for Russia and the United States but also for other countries faced with the problem of cleaning up sites contaminated with radioactive materials.

Overview Presentations

5

Ensuring Nuclear and Radiation Safety in the Use of Nuclear Energy for Peaceful Purposes

Andrei B. Malyshev, Russian Federal Atomic Energy Agency (Rosatom)

Environmental problems associated with the use of nuclear energy, including those that accumulated over the half-century of the arms race, concern both specialists and the public. It is obvious that the future of nuclear technologies both in the Russian Federation and in the world as a whole depend on their successful resolution.

The atomic industry, which began to take shape in the interest of defense in the 1940s and 1950s, proceeded along a complex path of development from scientific discoveries and technological inventions to subsequent industrial production and implementation. This was a difficult period, including the establishment of the nuclear industry and the development of the nuclear-powered submarine fleet, the sources of our main ecological problems today. Meanwhile, the use of nuclear energy for peaceful purposes also began.

The nuclear power sector developed so that in 2007, 10 nuclear power plants are operating (31 reactor units), with a total established capacity of 23.2 GW. This complex includes the following:

- Northwest Region: Leningrad plant, 4 GW; Kola plant, 1.8 GW
- Central Region: Smolensk plant, 3 GW; Novovoronezh plant, 1.8 GW; Kursk plant, 4 GW; Kalinin plant, 3 GW
- Southern Region: Volgodonsk plant, 1 GW

- Volga Region: Balakovo plant, 4 GW
- Urals Region: Beloyarsk plant, 0.6 GW
- Far East Region: Bilibino plant, 0.05 GW

In 2006 the nuclear power plants produced 154.6 billion kWh of electricity. They produced 16.5 percent of total electricity output (29.3 percent in the European portion of Russia) and comprised 11 percent of total established production capacity.

The nuclear industry has always devoted a great deal of attention to ensuring nuclear and radiation safety. This has been based on measures that have been generally adequate for existing technological capabilities and on the level of knowledge regarding the effect of radiation on human health and the environment. However, many of these measures are now unacceptable from the standpoint of current knowledge.

Today the nuclear and radiation security situation in the country is characterized by the presence of a number of facilities that do not meet modern safety requirements. The government recognizes the need to resolve accumulated problems and develop a system for ensuring and monitoring nuclear and radiation safety. In particular, a 2006 analysis shows a large volume of deferred problems in need of systematic and comprehensive efforts. Reactor units have been decommissioned at 5 nuclear power reactor sites, 13 industrial uranium-graphite reactor sites, 17 research reactor sites, and sites of various critical fuel assemblies and other types of equipment. A total of 198 atomic-powered submarines are being decommissioned from the naval fleet as follows:

- Dismantled and disposed of: 148
- Currently being dismantled: 23
- Awaiting dismantlement: 24
- Special resolution required for submarines in critical condition: 3

For a long time, such problems were addressed on an emergency, accident-response basis. Now, in the interest of dealing with these problems in an effective and coordinated manner, the Federal Targeted Program for Ensuring Nuclear and Radiation Safety for 2008 to 2015 has been developed and is in the process of being approved. The concept for this program was ratified by the government in April 2007.

Implementation of this program will make it possible to accomplish priority objectives in creating basic infrastructural elements needed to handle spent nuclear fuel and radioactive wastes and to eliminate many problems associated with past activities, including rehabilitation of radiation-contaminated facilities and sites. It is expected that resolution of the problems of our "nuclear heritage" will be financed from the state budget, while the processing of newly created spent fuel and radioactive wastes will be financed by market entities that must accumulate the necessary funds.

A key task is creation of a system for managing nuclear and radiation safety. There should be a logical concept based on the entire life cycle. An institutional model defined by the state and industry is necessary. Engineering infrastructure is needed along with a financing mechanism to allow for the accumulation of funds. The Federal Targeted Program is important to provide budgetary funds. Newly created spent fuel and radioactive waste must be taken care of.

Rehabilitation of radiation-contaminated sites is a mandatory element of the ongoing work of our enterprises, although it must be recognized that this work is now being carried out on a very limited basis. However, efforts to clean up such sites have taken on a broad scope in connection with the implementation of projects for the comprehensive dismantlement and disposal of decommissioned nuclear-powered submarines.

The next important step must be the implementation of a range of measures stipulated in the Federal Targeted Program for Ensuring Nuclear and Radiation Safety. In particular, of course, this includes efforts at the Mayak Production Association. In fact, the currently observed level of security for personnel, the public, and the environment with regard to the impact of the radiation-contaminated facilities at Mayak, including the industrial reservoirs, is generally at an acceptable level. That is, radiation effects on personnel, the public, and the environment are within currently existing norms. The exception, of course, is the impact on wildlife inhabiting the industrial reservoirs at the enterprise.

However, the risks caused by the problems that accumulated during the arms race could potentially entail significant damages. This primarily concerns problems of the industrial reservoirs operated by Mayak, including problems of the Techa Cascade reservoirs. The current reservoirs at Mayak are as follows:

(a) Techa Reservoir Cascade
- Area: 6,740 ha
- Volume: 358 million m^3
- Activity: 0.337 million Ci

(b) Reservoir V-9 ("Karachai")
- Area: 11 ha
- Volume: 0.4 million m^3
- Activity: 120 million Ci

(c) Reservoir V-17 ("Staroye Boloto")
- Area: 14 ha
- Volume: 0.3 million m^3
- Activity: 1.2 million Ci

Many problems associated with our nuclear heritage cannot be resolved without massive scientific support and consideration of international experience.

For many years, specialists from the Russian Academy of Sciences (RAS) have provided serious assistance in our work. It is necessary to mention the efforts of the RAS Institute of Geology of Ore Deposits, Petrography, Mineralogy, and Geochemistry at Lake Karachai and the Techa Cascade and in assessing the ecology surrounding uranium deposits. Also of importance has been the work of the RAS Nuclear Safety Institute on the comprehensive dismantlement of nuclear submarines, resolution of Mayak problems, and general coordination of research efforts aimed at ensuring nuclear and radiation security.

Many interesting and important results have been obtained as a result of cooperation between the Russian and U.S. Academies of Sciences and through work coordinated by the International Science and Technology Center. These efforts include work on radioactive waste management and the rehabilitation of contaminated sites. The practical experience of work done in the United States and other countries to decommission facilities involving nuclear and radiation dangers and to rehabilitate contaminated areas is very important to us.

The problem of rehabilitating contaminated sites has taken on added resonance in connection with the process of converting enterprises to joint-stock companies that we have begun in the industry. In our opinion the reform process that has begun within industry must not lead to a reduction in the level of environmental security that has already been achieved at enterprises, but must also promote resolution of accumulated problems. We have laid the foundation for this to a significant extent. This includes joint implementation of conceptually coordinated programs for the accelerated development of nuclear power and promotion of nuclear and radiation safety that have been developed, taking into account new forms of management and regulation regarding the use of nuclear power. Meanwhile, we need to shape a renewed policy of corporate management of environmental protection and utilization of natural resources. It is understandable that the foundations that have already been laid for environmentally safe operations must be securely consolidated and further developed.

We must remove the unfounded fears of the population associated with the activities of our enterprises. Therefore, information dissemination efforts must be an important part of our work. In recent years we have done a great deal in this regard. Industry reports on security are published openly each year, and a great amount of data on environmental protection issues has been posted on the Web sites of the Atomic Energy Agency.

Today's workshop has brought together leading specialists in the area of ensuring radiation safety. Their experience in cleaning up radiation-contaminated sites is of great scientific and practical interest. I believe that this workshop will be fruitful for all participants. I wish the participants successful work.

6

The Environmental Policy of the Russian Federal Atomic Energy Agency (Rosatom) and Priority Objectives for its Implementation*

Aleksandr M. Agapov, Rosatom, and
Leonid A. Bolshov, Russian Academy of Sciences (RAS) Nuclear Safety Institute

Literally before our eyes, issues of ecology and environmental protection have gone from general declarations and activities by enthusiasts to the ratification of modern norms under Russian and international law (see Box 6-1).

The path of sustainable development adopted by the global community entails not only political decisions but also partnership relations between the public and industry, the basis for which lies in the voluntary commitments of industry according to the values set forth in the Rio de Janeiro Declaration on Environment and Development (1992) and the Declaration of the World Summit on Sustainable Development held in Johannesburg in 2002. Coming to the forefront is the concept of environmental security, which is defined as a condition under which the natural environment and the vitally important interests of humanity are protected from possible negative effects of economic and other activities, extreme situations of a natural or technogenic nature, and their consequences. The vitally important interests of humanity are a broad concept, including high quality of life, the health of present and future generations, the presence of favorable environmental conditions, the preservation of biodiversity, the availability of natural resources for current and future generations, and so forth.

---

*Translated from the Russian by Kelly Robbins.

BOX 6-1
Development of the Concept of Environmental Security

1960-1975	Passage of the first environmental protection laws (United States, Western Europe)
1992	Rio de Janeiro Conference; concept of sustainable development
1996	First international standards, ISO series 14000
2002	Russian Federal Law on Environmental Protection

ENVIRONMENTAL SECURITY EFFORTS REGARDING THE RUSSIAN NUCLEAR INDUSTRY

More than any other industry both in Russia and worldwide, the nuclear industry senses the heightened interest of both the public and ecology specialists in ensuring environmental security. We often encounter harsh and not always constructive criticism. In this regard, this is an extremely appropriate time for an objective and open discussion of our achievements and unresolved problems.

The activities of our enterprises have traditionally been associated exclusively with radiation effects on people and nature. It should be noted that protecting people and the natural environment against radiation associated with the use of nuclear power is not a new task but rather an inalienable aspect of the activities of nuclear power enterprises over the course of many decades. Scientists and specialists from the State Research Center—Institute of Biophysics, the All-Russian Scientific Research Institute of Agricultural Radiology of the Russian Academy of Agricultural Sciences, and the Medical Radiological Research Center of the Russian Academy of Medical Sciences (RAMS) have made a significant contribution to ensuring radiation safety and resolving problems associated with radioecology and radiation medicine. We note that further steps toward strengthening the regulation of human radiation exposure limits must be weighed carefully. Furthermore, we believe that environmental security issues must be addressed comprehensively and systematically, taking into account all risk factors to which people today are subjected.

The task of ensuring environmental security has two limiting conditions. On the one hand, we, like all other industries, have limited resources. On the other hand, legislatively enhanced environmental limitations could lead to calls to halt the operation of certain facilities.

RELEVANT RUSSIAN LEGISLATION

Developing and implementing effective and practical environmental security solutions is a complex task. However, accomplishing this task is essential for further development of the industry, the prospects for which were clearly defined with the adoption of the Federal Targeted Program for the Development of the Russian Nuclear Power Industry Complex for 2007-2010 and up to 2015 (Government Resolution No. 605, October 6, 2006), which calls for accelerated development of nuclear power generating capacity in order to promote the country's energy security.

Russian President Vladimir Putin has repeatedly noted the undoubted need to resolve questions related to the safety of nuclear power and the nuclear industry in general. In late 2003, President Putin approved the Principles of State Policy on Ensuring Nuclear and Radiation Safety in the Russian Federation through 2010 (Pr-2196, December 4, 2003), which defines the goal, priority objectives, basic principles, and tasks of state policy on nuclear and radiation safety in Russia, as well as objectives for targeted program planning and management in this area.

A great deal of attention has traditionally been focused on environmental security in our industry. In 2003 the fundamental document entitled "Principles of the Environmental Policy of the Ministry of Atomic Energy (Minatom)" was developed and approved (ratified by Decree No. 67 of the Russian Minister of Atomic Energy, February 19, 2003). On April 5, 2005, a revised version of this document was approved by Decree No. 170 of the head of the Federal Atomic Energy Agency (Rosatom). Its current title is "Principles of the Environmental Policy of Rosatom." This document defines priority objectives and urgent measures to be taken in various areas. The next step must be the making of individual environmental protection commitments by enterprises, taking into account the specifics of their activities on the basis of the principles formulated in the Rosatom environmental policy document.

At this time, we can cite certain results of activities carried out on these objectives, including the following:

- Organizational and programmatic decisions of the Rosatom Board on the formation of departmental-targeted programs and radioactive waste management systems, problems of the Federal State Unitary Enterprise Mayak Production Association, development of an infrastructure for spent nuclear fuel management, comprehensive dismantlement of nuclear-powered submarines, decommissioning of facilities presenting nuclear and radiation hazards, and cooperation with the public
- Practical work by enterprises to develop and implement environmental protection measures
- Development and approval of initial special environmental programs
- Widespread initiation of international cooperation on comprehensive dismantlement and rehabilitation of shore technical bases at the Far Eastern Radioactive Waste Management Enterprise (DalRAO) and the Northern Radioactive Waste Management Enterprise (SevRAO)

- Expansion and augmentation of the scientific base of documentary evidence on the environmental safety of nuclear technologies
- Organizational decisions on improving the system for managing environmental security and nature protection activities

The agency's current structure and management functions are oriented towards efficiently achieving the strategic goals of Rosatom, one of the most fundamental of which is ensuring the safe and secure use of nuclear energy. Most of Rosatom's divisions deal to a varying extent with ensuring safety, coordinating the activities of enterprises under their control, or dealing with industry-wide matters.

Ensuring the safe operation of nuclear- and radiation-hazard facilities and precluding the possibility of serious accidents is a mandatory condition without adherence to which any considerations of the environmental safety of nuclear technology would be significantly devalued. No less important is the constant readiness of the industry to operate under extreme conditions. Qualitative advances have been made in this regard since the late 1990s.

The Scientific Coordination Council on Environmental Protection was created in 2005 to promote relevant scientific research and development of organizational and technical measures for the implementation of environmental policy. The council has been assigned the tasks of making scientific methodological recommendations and measures to develop the Industry-Wide System for Environmental Protection Management, as well as plans and practical measures to implement the Principles of the Environmental Policy of Rosatom. One effort that must be an important aspect of the council's activities is its effort to introduce at Rosatom enterprises national and international standards on environmental protection, rational resource use, environmental safety, and environmental management, primarily ISO series 14000 standards.

CURRENT ENVIRONMENTAL SAFETY SITUATION IN THE RUSSIAN NUCLEAR INDUSTRY

The present situation in the industry is characterized by Rosatom enterprises generally demonstrating a high level of environmental safety in accordance with the requirements of existing legislation. The main task in the industry is maintaining the level of environmental safety that has been achieved and constantly improving it in compliance with ISO standards. At the same time, problems have accumulated over decades with regard to spent nuclear fuel and radioactive wastes, the decommissioning of nuclear- and radiation-hazard facilities, and radioactive contamination of the environment.

The current nuclear and radiation safety situation in the country is characterized by three key factors:

1. The need to improve and develop state systems for ensuring and monitoring nuclear and radiation safety in the energy sector that are appropriate for the existing tasks and development plans in the nuclear power industry
2. The presence of facilities in the military-industrial complex that present nuclear and radiation hazards and do not meet current nuclear and radiation safety requirements (our "nuclear heritage") and that present a national security threat
3. The recognition of the need to resolve accumulated problems at the state level and the impermissibility of putting them off any further

In recent years, organizations have spent more than 20 billion rubles of their own funds annually to ensure the safety of facilities presenting nuclear and radiation hazards, including facilities in the military-industrial complex. However, the measures being taken are of an urgent, accident-response nature.

In order to address the accumulated problems in an effective and coordinated manner, the Federal Targeted Program for Ensuring Nuclear and Radiation Safety for 2008 to 2015 has been developed. The concept for this program was approved by Order No. 484-r of the government of the Russian Federation on April 19, 2007.

Implementation of this program will facilitate achievement of priority objectives in creating fundamental elements of the infrastructure to manage spent nuclear fuel and radioactive wastes and to eliminate problems associated with past activities, including the following:

- Removing from operation and (or) dismantling decommissioned facilities presenting nuclear and radiation hazards and ensuring that they are left in safe condition (including removal and reprocessing of spent nuclear fuel and radioactive wastes)
- Renovating facilities for the management of accumulated radioactive wastes and eliminating some repositories for solid and liquid radioactive wastes
- Decommissioning facilities of the Federal Agency for Marine and River Transport that present nuclear and radiation hazards and carrying out a set of efforts to manage spent nuclear fuel and radioactive wastes
- Removing accumulated volumes of spent nuclear fuel from research reactors and renovating systems for the physical protection of facilities presenting nuclear and radiation hazards
- Rehabilitating radiation-contaminated facilities and sites

A no less important condition for achieving the goal of nuclear and radiation safety lies in the reliable operation of systems for accounting, control, and physical protection of nuclear materials, radioactive substances, and radioactive wastes and in facilitation of the regime for nonproliferation of nuclear materials, prevention of unauthorized use of ionizing radiation sources, and monitoring of the radiation situation, exposure doses received by the population, and so forth.

DATA ON RESULTS OF NUCLEAR INDUSTRY ENVIRONMENTAL SAFETY EFFORTS

The nuclear industry needs to focus constant efforts on compiling scientific documentation regarding the level of safety that has been achieved, including that of environmental safety. For more than 30 years, issues related to environmental protection at industry enterprises have been analyzed and summarized by the industry department of environmental protection at the All-Russian Scientific Research Institute of Chemical Technology. On the whole, the environmental protection indicators of our production facilities are at a good level. Emissions and discharges of radionuclides are substantially lower than established legal limits. The exception is the Mayak Production Association—its problems require a special and comprehensive approach for resolution.

For many years, Rosatom has generally been achieving good indicators in its environmental protection activities. Nuclear industry enterprises contribute only about 0.5 percent of all industrial emissions of chemical contaminants and about 3 percent of all discharges of polluted wastewater. This is one of the best figures among all industrial sectors (only the communications industry is better).

An important task is developing and improving the base of scientific evidence on environmental security matters. Efforts to produce a comprehensive analysis of environmental risks in regions where our enterprises are located are extremely useful in this regard. Including factors of a nonradiation-related nature in this risk factor analysis will make it possible to determine real environmental protection priorities at the regional level. Interesting and important work in this area has been done by scientists from the RAS Nuclear Safety Institute, the Russian Research Center—Kurchatov Institute, and the State Science Center—Institute of Biophysics.

Research on the comparative analysis of radiation and chemical risks conducted at the RAS Nuclear Safety Institute in cooperation with the RAMS A.N. Sysin Scientific Research Institute of Human Ecology and Environmental Hygiene has shown that nuclear technologies have an extremely insignificant impact on the health of the population. In regions where nuclear industry enterprises are located, radiation risks are at a level of 10^{-7}-10^{-6}, and in regions where nuclear power plants are located these risks are 10^{-7}-10^{-8} (see Figure 6-1).[1] These levels are two to four orders of magnitude lower than the risks associated with chemical pollution of the environment.

As a comparison, the risks to the health of the population associated with the operations of a coal-fired power plant are up to 10^{-3} annually (see Figure 6-2).

---

[1]Risks in this paper are the annual risks of cancer to an individual. These are the acceptable excess upper bound annual cancer risks to an individual. Acceptable exposure levels for known or suspected carcinogens are generally concentration levels that represent an excess upper bound lifetime cancer risk to an individual of between 10^{-4} and 10^{-6} using information on the relationship between dose and response. In the international literature, these risks are usually presented as lifetime risks.—Ed.

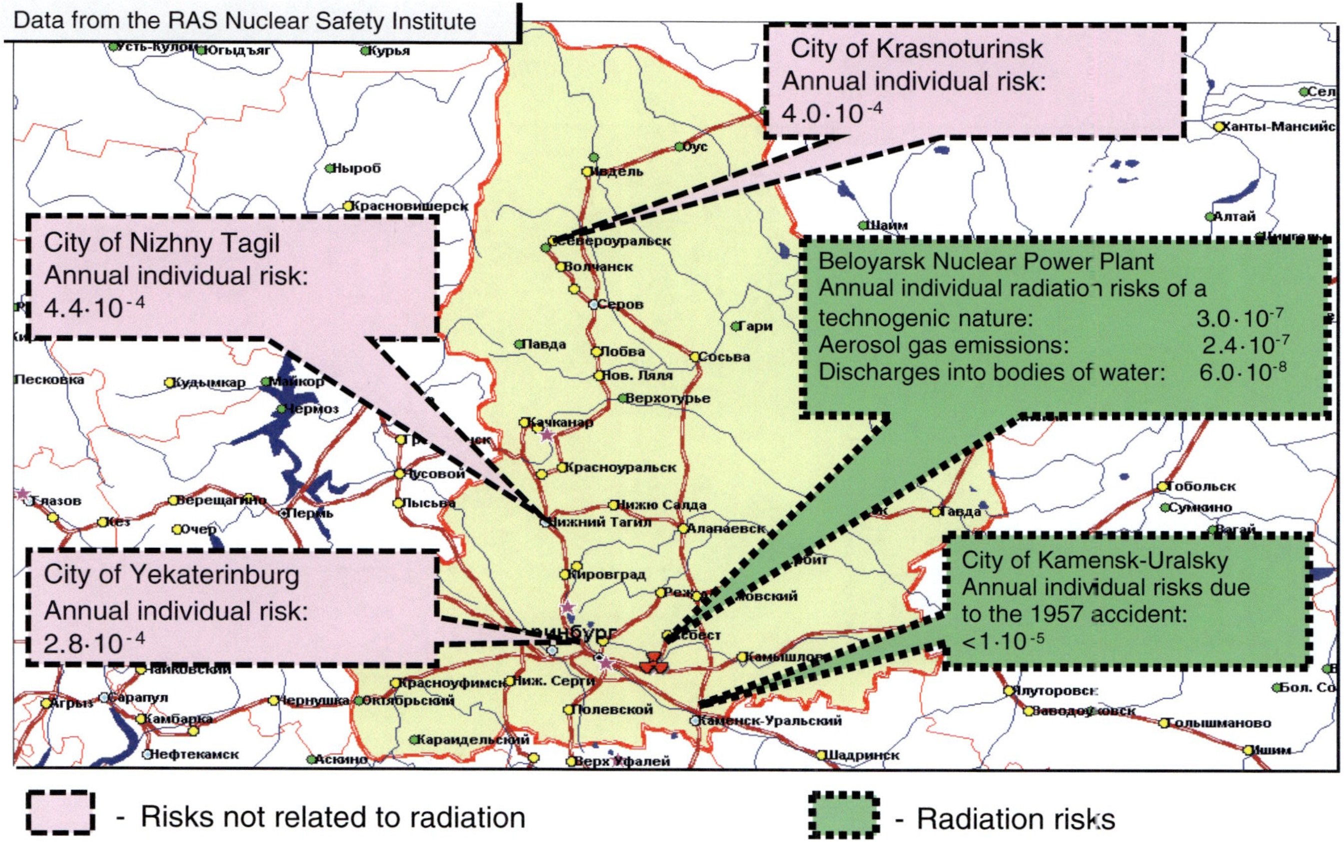

FIGURE 6-1 Regions of assessed risk in Sverdlovsk Oblast.

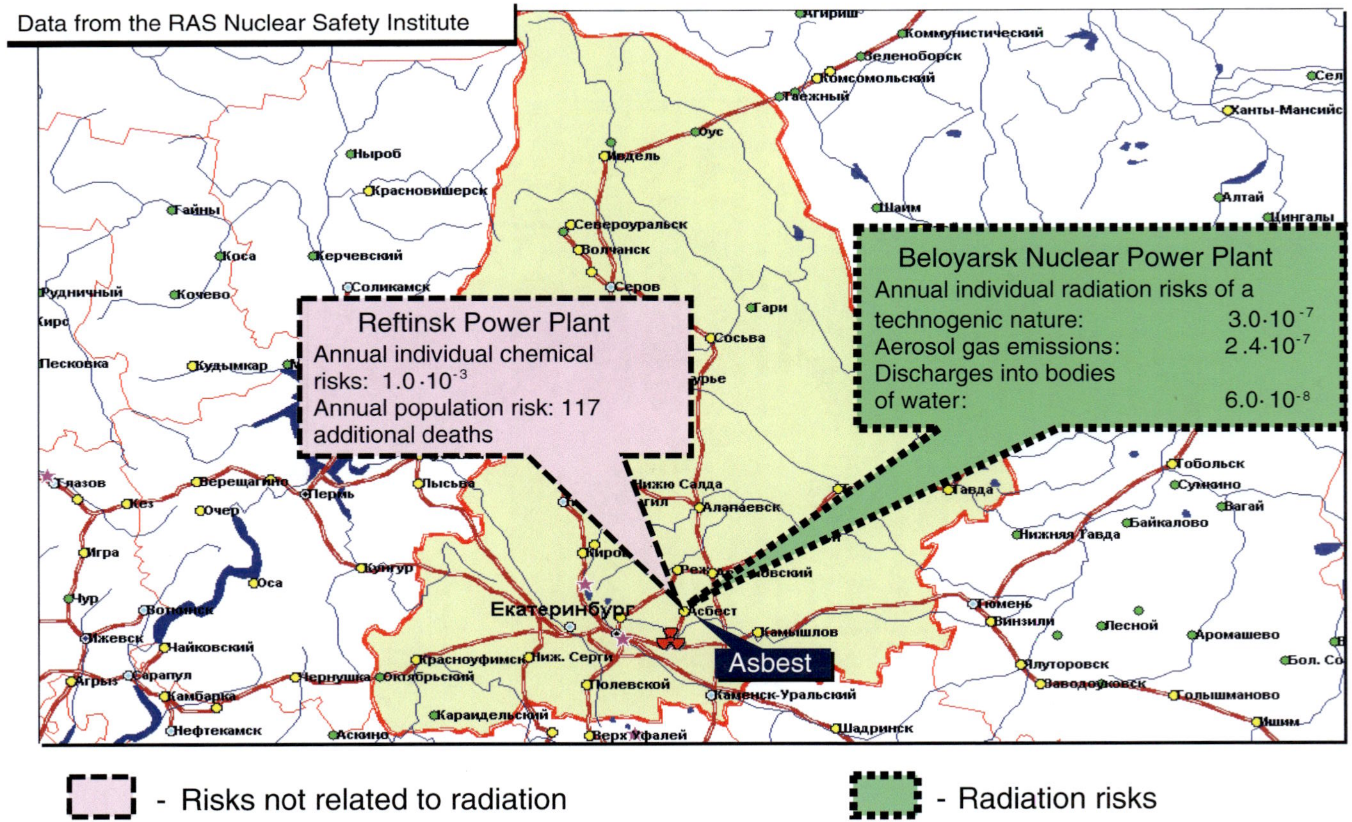

FIGURE 6-2 Risks to the health of the population living in regions near nuclear and coal-fired power plants in Sverdlovsk Oblast.

Among the industry's accumulated and not yet fully resolved problems are radioactive wastes and, in particular, the problems of the Mayak Production Association. However, positive strides have been made even at this very problematic enterprise. In 2005-2007 a wide range of work has been carried out to ensure the safety of the industrial reservoirs at Mayak. We note that on the initiative of Rosatom and within the framework of existing legislation, an interagency group of specialists has drafted a "Technical Regulation on the Safety of the Techa Reservoir Cascade." Further development of this draft regulation will make it possible to put in place a much-needed legitimate regulatory-legal basis for the safe operation of this reservoir cascade.

CREATION OF A UNIFIED STATE SYSTEM FOR RADIOACTIVE WASTE MANAGEMENT

Creating an effective radioactive waste management system is a fundamental condition for the further development of nuclear power and the nuclear industry in general (see Box 6-2). At present, the need has arisen to create a unified state system for radioactive waste management based on the organizations of Rosatom and under the agency's control. Such a system will make it possible to ensure safer management of radioactive wastes at all stages in the life cycle.

An important step in this effort must be the adoption of a radioactive waste management doctrine that would create a scientifically based and clearly defined vision for securing radioactive wastes over the long term. Significant efforts are

BOX 6-2
From the Decisions of the Rosatom Board "On the Formation of a State System for Radioactive Waste Management," April 6, 2005

- Prepare for presentation to the Government of the Russian Federation recommendations on working with other interested federal executive branch agencies the Russian government acts on

  - Affirming the doctrine on radioactive waste management in the Russian Federation
  - Defining a list of facilities for the creation of regional sites for the long-term storage and ultimate isolation of radioactive wastes
  - Identifying sources and procedures for financing radioactive waste management efforts

- Organize a Russia-wide inventory of radioactive wastes as of December 31, 2005

required to develop recommendations on organizational and financial mechanisms for the operation of a unified state system covering radioactive waste management activities.

The problems of the nuclear heritage must be addressed comprehensively and at the state level. However, the realities of life are such that we have dealt with them primarily independently, so as to facilitate ongoing enterprise operations and prospects for their development. The recent change and even the breakthrough in terms of state attention to the problems of the nuclear heritage are largely associated with the adoption and implementation of the Federal Targeted Program for Ensuring Nuclear and Radiation Security for 2008 to 2015.

THE SPECIAL PROBLEMS OF NUCLEAR-POWERED SUBMARINES

Issues regarding nuclear and radiation safety associated with the comprehensive dismantlement of nuclear-powered submarines hold a special position. This is a major activity directly aimed at ensuring long-term environmental security.

Recent and previous years have been characterized by the rapid expansion of international cooperation under both the Global Partnership and numerous bilateral agreements. In the aim of coordinating work being carried out under the framework of international cooperation, the first Strategic Master Plan for the Comprehensive Dismantlement of Nuclear-Powered Submarines in Northwestern Russia—a plan of urgent measures—was prepared in 2004 through joint efforts with the European Bank for Reconstruction and Development. In 2007, plans are being completed for the second stage of work, SMP-2. Led by RAS Academician A. A. Sarkisov and RAS Corresponding Member L. A. Bolshov, these wide-ranging and complex efforts involve the active participation of scientists from the RAS Nuclear Safety Institute, the Russian Research Center—Kurchatov Institute, and the N. A. Dollezhal Research and Development Institute of Power Engineering (NIKIET), as well as specialists from the enterprises of other interested agencies.

In 1998 the government of the Russian Federation assigned the task of comprehensive dismantlement of nuclear-powered submarines to the Russian Ministry of Atomic Energy. Table 6-1 presents information on the current status of such submarines removed from naval service.

Plans call for dismantling all nuclear-powered submarines removed from military service by 2010. More than 50 billion rubles will be required to finance programmatic measures under the Strategic Master Plan through 2010 (see the list of tasks in Box 6-3). It is expected that more than 30 billion rubles of the funds required will come from international technical assistance.

ISSUES ASSOCIATED WITH NUCLEAR POWER PRODUCTION

Ensuring environmental security associated with the use of nuclear power is an important element in Rosatom's system of strategic priorities. It is one of the

TABLE 6-1 Data on the Status of Nuclear-Powered Submarines Removed from Naval Service as of February 1, 2007

Submarines	Total	Northern Region	Pacific Region
Removed from service	198	120	78
Dismantled	148	97	51
Currently being dismantled	23	10	13
Awaiting dismantlement	24	12	12
Special resolution required (submarines in critical condition)	3	1	2

fundamental conditions for the sustainable development of the nuclear industry in the twenty-first century. The limited nature of hydrocarbon fuel resources raises the question of how to meet the world's energy demands. Given the shortage of cheap natural gas in the country and the negative environmental impacts of coal-fired power plants, developing nuclear power under the condition that we shift to a closed nuclear fuel cycle (which will ensure nuclear fuel supplies for thousands of years) is the most environmentally safe way to ensure a sustainable energy supply for Russia. As a point of information, the production of power at nuclear power plants saves more than 40 billion m^3 of natural gas each year.

Nevertheless, enterprises and Rosatom inevitably must address complex problems of maintaining and consolidating their positions on the Russian and world markets for energy technologies and nuclear technologies. All competitive mechanisms are in play when commercial projects are carried out, including relying on public opinion on matters of the ecological impacts of products and

BOX 6-3
Comprehensive Dismantlement of Nuclear-Powered Submarines: Tasks through 2010

- Dismantle 83 submarines, 2 surface ships with nuclear power blocs, and 16 nuclear technical service ships
- Unload, transport, and reprocess 100 t of spent nuclear fuel
- Safely maintain all comprehensive dismantlement sites
- Isolate 2 submarines in critical condition in Primorsky Krai
- Build facilities for long-term onshore storage: 190 reactor section blocs in Murmansk Oblast and Primorsky Krai
- Reprocess 30,000 t of solid radioactive wastes and 12,000 m^3 of liquid radioactive wastes
- Rehabilitate spent fuel and radioactive waste repositories as well as contaminated sites at 3 shore technical bases

the processes for their production. Along with the basic positive content involved, ecology, like "democratization" and "human rights," has become a powerful tool to be used in the competitive struggle of local elites, various energy technologies, countries, and groups of countries.

The clearest example is the Kyoto Protocol, which excludes nuclear energy from the flexibility mechanisms, despite its absolutely obvious potential role in reducing greenhouse gas emissions. This artifact must be corrected. Meanwhile, this does not hinder the active use of the environmental advantages of nuclear power in Russia. According to Article 2 of the protocol, policies and measures aimed at allowing each signatory party to fulfill its protocol obligations are developed and implemented independently by each party in accordance with its own national conditions. Further development of nuclear power in Russia will promote a reduction in greenhouse gas emissions and increased opportunities for trade in emissions quotas. In this regard, it is essential to establish Rosatom's policy on the right of the nuclear power industry to a share of the Russian quota, as well as rules for the trade in quotas on the domestic market.

THE CONCEPT OF ENVIRONMENTAL QUALITY MANAGEMENT

A characteristic feature of the modern approach to ensuring environmental security is the management of environmental quality (see Figure 6-3). Modern environmental quality management systems grew out of quality assurance systems and have become integral and effective tools regulated by international standards such as the ISO-14001 series and others. They currently offer a wide range of proven solutions and procedures. In particular, a strategic approach to ensuring environmental security at enterprises is implemented by means of risk assessment and risk management both under normal operations and when accidents occur.

In recent years, the industry has substantially stepped up its efforts to train specialists on the requirements of ISO-14000 standards. Such courses are conducted by the Atomenergo Interagency Institute for Qualification Improvement in cooperation with the RAS Nuclear Safety Institute and the State Central Institute for Qualification Improvement.

The experience of the Nuclear Safety Institute in organizing (jointly with the U.S. Department of Energy) the planning and presentation of courses to introduce ISO-14001 series standards has shown that there is real interest at Rosatom enterprises in establishing modern systems for enterprise management and certification, as well as real potential for their implementation.

Among the enterprises in the industry, there are leaders that have already created modern environmental management systems. A number of enterprises have implemented systems meeting international ISO-14000 standards and have certified them with independent certification organizations. These include the fol-

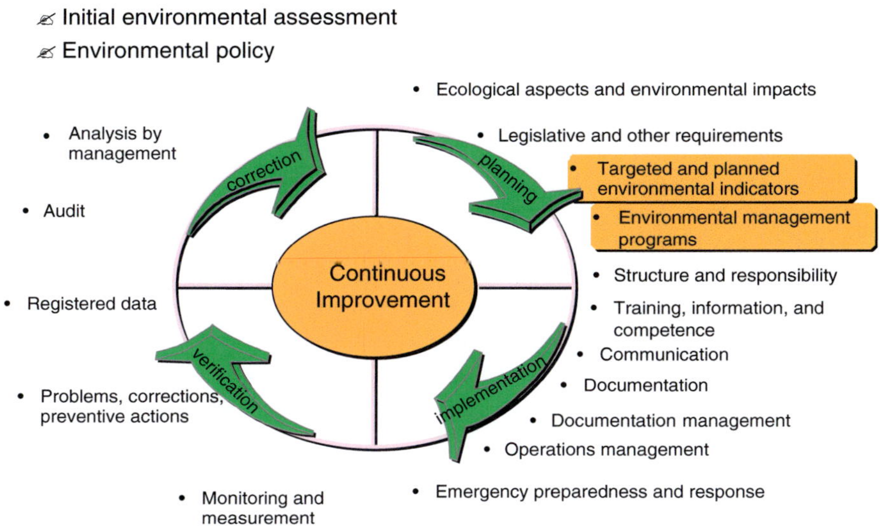

FIGURE 6-3 Model environmental management system.

lowing enterprises of the TVEL Corporation: the Chepetsk Mechanical Plant, the Novosibirsk Chemical Concentrates Plant, and the Machine-Building Plant.

The Rosenergoatom Concern is making active efforts in this regard, having developed and adopted a Program of Work on Environmental Management Systems Certification at the concern and its subsidiary nuclear power plants in accordance with ISO-14000 requirements. In early 2005 the Balakovo Nuclear Power Plant received the first certificate of compliance. There are also problematic enterprises in the industry, primarily those that are "hostages" to serious past problems; however, all must move forward on addressing current and inherited environmental problems.

In noting the initial successes on introduction of the ISO-14001 standards, it must be mentioned that an environmental management system is a process, not an event. An effective system is a "living" entity, evaluating and correcting activities and aimed at their constant improvement.

ISO-14001 series standards call for the formulation and approval of enterprise environmental policies. The adoption of the Principles of the Environmental Policy of the Russian Ministry of Atomic Energy (2003) and its new version, Principles of the Environmental Policy of Rosatom (2005), represented a correct and important step in ensuring the development of the industry. In particular, practical implementation of environmental policy will make it possible to avoid situations associated with new and, on the world market, mandatory requirements, which if not met would undermine the competitiveness of the products of

Russian nuclear fuel cycle enterprises. It is characteristic and completely natural that the export-oriented enterprises of the TVEL Corporation and the Rosenergoatom Concern are among the leaders in instituting modern environmental management systems. In their environmental policy statements they clearly highlight the priority placed on ensuring security and protecting the environment.

It would be expedient to begin preparing Rosatom enterprises for certification to appropriate international standards now, without waiting for the completion of the industry restructuring process. To facilitate the procedure for certifying environmental management plans at individual enterprises, a packet of materials on standards from the current series has been prepared, along with practical recommendations on implementation of certification procedures.

INFORMATION POLICY AND PUBLIC PERCEPTIONS

Information policy on environmental matters is especially important for the further development of the industry. We must work to ensure that public perception of the environmental impacts of our enterprises corresponds with reality. Having convincing proof and scientific evidence of environmental safety in the eyes of the public, including at the level of the average taxpayer and voter, is no less important than actually ensuring environmental security itself. In very recent history there have been a significant number of examples of the effects of negative public opinion regarding nuclear activities in the form of rejections of nuclear power at the state level.

Public reports on environmental matters are an effective type of information activity. For this reason, we annually publish open industry reports on safety that present the current situation, existing problems, and environmental protection measures taken; analyze activities; and cover the search for new ways of ensuring all aspects of environmental security. We see such types of voluntarily published reports as a means of demonstrating to a broad range of interested parties our practical adherence to principles of openness, environmentally safe operations, and sustainable development.

The Web sites of Rosatom, the Rosenergoatom Concern, and other enterprises present a great volume of data on natural resource utilization and environmental protection issues. However, simply presenting information alone does not make it possible to achieve the agency's ultimate information policy goals. We need even more active work by enterprises, cooperation with the regions, and informational materials designed for the general public. We must learn to operate effectively under information crisis conditions and in the face of provocations.

The recent public affairs-related events surrounding the Rostov Nuclear Power Plant (specifically, the events of Sunday, May 20, 2007), when the population was alarmed by false information circulated in the region about an emergency release at the plant, showed that we are still unprepared to react quickly to such provocations. This is despite our industry, more than any other, understanding the danger of such informational attacks and the scope of their associated socio-psychological consequences for the public. In regions where enterprises using nuclear technologies are located, the creation of information centers capable of online interaction with the population should obviously be viewed as an urgent task for the industry with regard to the public affairs aspects of environmental security.

FUTURE PROSPECTS FOR THE RUSSIAN NUCLEAR INDUSTRY

Regarding future prospects, in our opinion the process that has begun in the industry to convert enterprises to joint-stock companies must not lead to a reduction in the level of environmental security that has been achieved. We expect that implementation of programs to develop the nuclear energy industry complex and develop nuclear and radiation safety, which have been created taking new forms of management into account, will promote resolution of our accumulated environmental problems. Within the framework of corporate management of the use of nuclear power, we must work to strengthen and further develop the foundations that have already been laid for environmental security efforts.

The most important environmental security-related tasks we face at the federal level include the following:

- Forming a state system for ensuring and monitoring nuclear and radiation safety associated with the use of nuclear power
- Creating an effective state system for radioactive waste management
- Creating effective mechanisms for the technical regulation of environmental security
- Developing the system of state environmental security guarantees, including insurance against nuclear damages
- Improving the system of environmental protection management
- Promoting outstanding development of scientific potential in the area of environmental security, which is necessary to provide scientific foundations for activities in this sphere

The top priority tasks for Rosatom enterprises must be the following:

- Unconditionally meeting the requirements of environmental protection legislation
- Promoting the evolving increase in levels of environmental security by introducing modern environmental management systems and effective new technologies

These tasks may be accomplished successfully only if the industry collaborates closely with government agencies and the public at all levels.

7

Evaluation of Radiation Ecology Status Around Russian Nuclear and Radiation Enterprises Based on Landscape-Geochemical Research*

V. I. Velichkin, Ye. N. Borisenko, A. Yu. Miroshnikov, V. I. Myskin, N. V. Kuzmenkova, and I. I. Chudnyavtseva, Russian Academy of Sciences Institute of Geology of Ore Deposits, Petrography, Mineralogy, and Geochemistry

INTRODUCTION

Past experience in radiogeoecological work aimed at rehabilitating radiation-contaminated areas indicates that the effectiveness of such rehabilitation measures largely depends on the reliability and quality of the initial scientific foundations on which the measures were developed and based. The data traditionally presented as the basis for rehabilitation of contaminated areas include information on the composition, quantity, and distribution of contaminants and on the soil cover in which they are found. This information makes possible an overall assessment of the current radiochemical status of the environment at the sites being studied, but it is insufficient both to reveal the conditions and mechanisms of contaminant migration and concentration and to predict how the studied ecosystems will change over time and space.

Since 1990, at the initiative and under the leadership of Academician N. P. Laverov, the Russian Academy of Sciences Institute of Geology of Ore Depos-

*Translated from the Russian by Kelly Robbins.

its, Petrography, Mineralogy, and Geochemistry (IGEM) has been conducting landscape-geochemical research scientifically based on the teachings of Professor A. I. Perelman and his students regarding geochemical landscapes and geochemical barriers. The techniques developed at IGEM especially for assessing the environmental impacts of nuclear and radiation facilities have been elaborated and applied in practice by specialists from the institute's Laboratory of Radiogeology and Radiogeoecology. These techniques include the following:

- Detection and characterization of actual and potential radiation sources
- Identification and integrated study of the hierarchy of landscape geochemical systems as well as elementary landscapes and their transition zones
- Studies within each landscape type of the composition and geochemical parameters of soil, vegetation, substrate rocks, surface water and groundwater, and bottom sediments of swamps, lakes, rivers, and reservoirs
- Studies and assessments within each elementary landscape of distribution and availability modes of radioactive and stable contaminants in soil, vegetation, substrate horizons, river and lake water, and bottom sediments
- Integrated assessment of migration and accumulation conditions of radioactive and stable contaminants within landscape geochemical systems and elementary landscapes on the basis of their geomorphological, soil-lithological, geochemical, hydrogeological, and biological features
- Disclosure and studies of natural geochemical barriers and formation conditions of technogenic[1] geochemical barriers
- Creation of landscape geochemical and ecological geochemical maps in various scales

Cartography is an integral component of these techniques. Maps provide the necessary basis for detecting areal diversity of conditions for the distribution of radioactive substances, for assessing and predicting changes in the radioecological environment over space and time, and for supporting decisions on optimizing environmental management.

This research methodology makes it possible to determine the characteristics of transfer and accumulation of radioactive and stable contaminants, and on this basis, to predict more reliably the ecological status and further changes in natural landscapes in regions where nuclear and radiation facilities are located. IGEM specialists are using this methodology to study a number of major nuclear and radiation sites in Russia located in various climatic zones. This paper focuses on results of the first stage of this research.

---

[1]*Technogenic* is used to refer to phenomena arising as a result of the development or deployment of technology.

SOURCES OF RADIATION AND CHEMICAL CONTAMINATION AT RUSSIAN NUCLEAR AND RADIATION ENTERPRISES

This paper covers three groups of major Russian nuclear and radiation enterprises. The first includes the Mayak Production Association radiochemical complex, a multipurpose dual-use facility that in its initial phase of activity was oriented exclusively to work on defense-related nuclear programs, but was later converted to focus on eliminating the consequences of defense-related activities and addressing various problems associated with the nuclear power fuel cycle. The second group includes the Nerpa Ship Repair Plant, which is among the enterprises that carry out specific operations connected with eliminating nuclear weapons carriers. The third group includes facilities engaged in work on particular stages of the nuclear fuel cycle, such as uranium ore extraction and processing enterprises (the Priargunsk Mining and Chemical Production Association—PMCPA) and electricity producers (the Smolensk and Balakovo nuclear power plants) (see Figure 7-1).

The bulk of the radionuclides contaminating the environment at the nuclear and radiation enterprises studied are technogenic in nature; that is, they were created in the course of processes occurring in nuclear reactors. Before the initiation of work in the nuclear energy field, most of these radioisotopes did not exist in nature. Naturally occurring radioactive elements are known ecosphere contaminants only at the PMCPA uranium mining complex, which is located in the eastern part of the Lake Baikal region.

FIGURE 7-1 Locations of major Russian nuclear and radiation enterprises. 1—Nerpa Ship Repair Plant; 2—Smolensk Nuclear Power Plant; 3—Balakovo Nuclear Power Plant; 4—Mayak Producation Association; 5—Priargunsk Mining and Chemical Production Association.

The contamination sources, their toxic substance content, and the mechanisms of their spread into the environment are characterized as follows at each of the nuclear and radiation enterprises studied.

The Mayak Production Association, Russia's oldest and largest radiochemical enterprise, has the largest number of known sources of radioactive contamination. In the course of Mayak's 60 years of operations, two periods may be singled out during which there were various active sources of radiation contamination. In the initial period (the 1950s and early 1960s), there were uncontrolled gas-aerosol emissions of radioactive substances into the atmosphere and discharges of liquid wastes, including highly active wastes, into the Techa River. From 1949 through 1956, this river received discharges of liquid radioactive wastes with a total activity level of 2.5 million Ci, which led to intensive radiation contamination of silt deposits in the Techa River and of the soil layer in its valley. The same period experienced wide-scale environmental contamination due to accidents. First, a container of highly radioactive liquid wastes exploded, subsequently discharging into the atmosphere finely dispersed radionuclides with a total activity of about 2 million Ci. Later, the winds spread a radioactive cloud of drought-created dust from the shores of Lake Karachai, an accumulation reservoir for liquid radioactive wastes. These accidents resulted in the formation of the East Urals Radioactive Trace, which has an area of more than 1,500 km^2.

At present, the currently operating enterprises of the Mayak Production Association emit into the atmosphere a regulated quantity of radioactive aerosols that causes practically no environmental damage. Planned discharges of low-activity liquid radioactive wastes are also made into the reservoirs of the Techa Cascade and of medium-activity liquid wastes into Lake Karachai. These releases result in the radiation contamination of the groundwater lying beneath these bodies of surface water.

The primary radionuclides in the discharged gas-aerosol fractions and liquid radioactive wastes are short-lived plutonium-106, zirconium-95 + niobium-95, cesium-144, cobalt-60, zinc-65, and tritium; medium-lived cesium-137 and strontium-90; and a significantly smaller quantity of long-lived uranium-238 and transuranic elements (plutonium, neptunium, americium, and curium). Except for the short-lived ones, all the remaining radionuclides are found in various quantities in previously contaminated soils, rock, and surface water and groundwater.

At the Nerpa Ship Repair Plant, which dismantles decommissioned nuclear-powered submarines, the main operations include cutting up the submarine hulls and reprocessing their internal parts. These operations result in the formation of finely dispersed mechanical particulate clouds containing radioisotopes (cesium-137, strontium-90, and cobalt-60) and heavy metals (iron, copper, nickel, titanium, and others) that are spread by air currents into the environment.

During their routine daily operations, the Smolensk and Balakovo nuclear power plants regularly carry out controlled air emissions containing radioactive inert gases and aerosols of the radionuclides cesium-137, cesium-134, strontium-

90, cobalt-60, manganese-54, carbon-14, and iodine-131, as well as transuranic elements. The concentrations of these emission components are substantially lower than maximum allowable levels. However, their progressive accumulations in soils could exceed allowable standards over time.

Located near the city of Krasnokamensk in the eastern part of the Lake Baikal region, the PMCPA is the only enterprise in Russia that mines and processes molybdenum-uranium ores from the Streltsov Ore Field, which is unique in the size of its total uranium reserves. Here the environmental contaminants include natural radionuclides (uranium-238, uranium-235, and their decay products radium-226 and radon-222), thorium-232, potassium-40, and stable elements (molybdenum, manganese, arsenic, and beryllium), as well as sulfuric, nitric, and other acids. The contamination sources include tailings repositories for uranium ore processing and sulfuric acid production plants; dump sites for poor and unbalanced ores for heap leaching of uranium; an ash repository at the heating plant, which uses local hard coal; discharge sites for mine water; dump sites for unbalanced ores and worked rock; areas where uranium ores are strip mined; entrances to underground mines; and roads on which ore materials and worked rock are transported.

Contaminants are transferred from the soil surface in the form of fine particle suspensions transported by surface water or atmospheric currents. In surface water the toxic substances migrate in suspended and solute form, while in groundwater they circulate in solutions and colloidal suspensions.

GEOCHEMICAL LANDSCAPES AND THEIR RADIOECOLOGICAL STATUS IN REGIONS WHERE NUCLEAR AND RADIATION ENTERPRISES ARE LOCATED

As further characterized below, regions with nuclear and radiation enterprises are located in all of the landscape-geochemical zones of Russia: tundra (the Nerpa Ship Repair Plant), forest (Smolensk Nuclear Power Plant), forest-steppe (Mayak Production Association), steppe (Balakovo Nuclear Power Plant), and dry steppe (PMCPA). This makes it possible to carry out a comparative analysis of conditions for the migration and concentration of natural and technogenic radionuclides in various landscape-geochemical settings.

The Nerpa Ship Repair Plant specializes in dismantling decommissioned nuclear-powered submarines. This enterprise is located in the northern part of the Kola Peninsula, in the tundra landscape zone. The natural landscapes of this area, called "Murmansk-like," formed on dense, poorly permeable, crystalline Precambrian rock, creating a topography of plains and folded rocky uplifts and depressions. On the basis of these rocks and because of their specific physical characteristics, primitive shallow soils were formed (to the depth of a few tens of centimeters). These soils may be characterized as mildly acidic, organogenic, highly peaty, hydromorphic formations with moderately restorative reactivity.

They possess a heightened capacity for accumulating radioactive and stable contaminants. Water migration of dissolved forms of these elements is complicated in these soils. The contaminants are transferred mainly in a suspended particulate state in the atmosphere and in surface water flows.

Autonomous and subsidiary elementary geochemical landscapes have been defined in the landscape-geochemical system around the Nerpa Ship Repair Plant.[2] It has been established that in the soil layer of the autonomous landscapes, the highest average contents of radioactive cesium (97 and 77 Bq/kg) are found in soils at the industrial site and in the 15 km zone surrounding it. Meanwhile, these indicators reach only 39 Bq/kg in the 50 km zone around the plant. It is most likely that average cesium-137 content levels in the soils in this zone are due to global atmospheric fallout and may appropriately be viewed as background levels in the area of the Nerpa plant, as at the plant site and in the 15 km zone, in addition to background levels, there is the cesium-137 that, as previously noted, is carried by air currents from the site where the submarines are dismantled. Typically, from 79 to 92 percent of the total quantity of radioactive cesium accumulated in soils is concentrated in the top 3-6 cm layer, which is the most enriched with organic compounds. Increased concentrations of heavy metals have also been discovered in soils at the industrial site and in the 15 km zone.

Thus, the environmental impact zone for the activities of the Nerpa Ship Repair Plant was determined in the course of the research. The increased levels of radioactive cesium and heavy metals found in the zone are significantly lower than maximum allowable concentrations. Nevertheless, the radioecological status of the environment around the Nerpa plant requires systematic monitoring.

The Smolensk Nuclear Power Plant is located in the forest landscape zone of Central Russia. The forest and agricultural landscapes of this region formed on sandy-clay, weakly fractured rock from the old East European Platform, with topographic conditions characterized by gently sloping hills and plains. A significant part of the area is forested, with low-lying parts of the plains and riparian floodplains being swampy. Weakly mineralized with hydrocarbonates and calcium, the groundwater lies at a depth of 30 cm to up to 3 (and rarely 10) m. The soils are sod-podzolic and podzolic-gley in sands, loamy sands, light loams, and clays, as well as meadow, peat-gley, and peaty.

Two types of geochemical landscapes may be singled out: (1) Polessk-type, which are flat, swampy, and forested plains on sands and loamy sands; and (2) Smolensk-type, which are hilly, lightly forested, raised plains and eroded uplifts on clay and clay-sand moraine deposits.

Polessk-type landscapes are more hydromorphic. They are characterized by processes of leaching of radionuclides from the topmost soil layers, low sorption

---

[2]Autonomous landscapes are parcels of the studied areas with a higher topographic elevation. Subsidiary landscapes are parcels with less elevated relief, which often contain river and stream channels or lake depressions.

capacity of geochemical barriers within the soil, and weak contaminant fixation, which on the whole promotes their vertical and lateral migration. However, because of the low contrast between the autonomous and superaqueous landscapes, this migration proceeds slowly. Radionuclides have accumulated primarily in the peripheral zones of swamps in the alkaline gley sorption geochemical barrier.

Smolensk-type landscapes are less swampy. The leaching of radionuclides from the topmost layers of sod-podzolic, loamy, and clay soils is less pronounced than in the Polessk-type areas. The geochemical barriers within the soil have a higher sorption capacity and limit the vertical migration of radionuclides. These landscapes are also characterized by processes of sheet erosion and accumulation of radionuclides with soil particles less than 1 mm in diameter from the foothill slopes, eolian dispersal, and displacement of contaminants in the soil layer during tillage.

It has been established that radionuclides generated by the Smolensk Nuclear Power Plant during normal, accident-free operations enter the environment in small quantities and are localized in soils very close to the plant.

Radionuclides are absorbed by sediments at the bottom of the reservoir and by humus and peat layers in the soils and precipitate out on the periphery zones of swamps, which represent a complex geochemical barrier of absorbent gley and alkaline components. This is the most effective concentrator of substances in the landscape zones affected by the Smolensk plant.

The research that has been conducted indicates that the overall radioecological situation in the area around the Smolensk Nuclear Power Plant meets current safety requirements. The radionuclides emitted by the plant during its normal operating regime are fully contained by the natural geochemical barriers that exist in the landscape.

The Balakovo Nuclear Power Plant and a large portion of the 30 km zone around it are located in the steppe landscape zone of Central Russia (the Syrtovaya Plain of the Volga area around Saratov). The landscapes of the black-earth steppe that developed in this area were formed on carbonate-sand-gley weakly violated rock on the sedimentary mantle of the ancient East European Platform. The topography is predominantly marked by flat plains divided by a network of infrequent gullies into separate flat, steeply sloped uplifts where water exchange processes are greatly slowed. The autonomous and subsidiary landscapes exhibit little lithological and geochemical contrast.

The soils are chernozems with high and medium humus content and varying degrees of carbonate content, lixiviation, and solonetz characteristics. They are characterized by calcium-class water migration, and groundwater reactivity is neutral, slightly alkaline, and rarely slightly acidic. Chernozem soils possess increased capacity for irreversible absorption of cesium-137. In essence, the top layer of chernozem soils is an active spatial geochemical barrier to radioactive cesium. At the same time, wind and mechanical transport processes are fairly widely developed.

Analysis of the quantitative aspects of cesium-137 distribution in the soil layer within the 30 km zone indicated that the overall level of cesium-137 content (no more than 40 Bq/kg) in the soils of the region studied is substantially lower than allowable standards. Furthermore, more than 90 percent of this radioisotope was concentrated in the top 3 cm of the soils, the layer most enriched with organic matter.

The data gathered make it possible to conclude that small quantities of radionuclides deposited daily due to regular atmospheric emissions by the Balakovo Nuclear Power Plant are firmly fixed and progressively absorbed by the soils. This tendency means that regular monitoring observations must be conducted in the area around the Balakovo plant.

The Mayak Production Association is located in a forest-steppe landscape zone and is geographically associated with the eastern slope of the southern segment of the Ural mountain range. The western part of the region lies in the Ural foothills, while the eastern part lies in the hilly plain of the West Siberian Platform plate. The forest-steppe landscapes of this area were formed on intensively dislocated mica shales (in the western part of the region), on sedimentary-volcanic andesite-basalt rock (in the central part), and sand-clay-carbonate platform deposits (in the eastern part).

The special characteristics of the topography mean that the region is divided into strongly predominant (about 75 percent of the area) autonomous (alluvial) landscapes in the elevated sections and the subsidiary (superaqueous) landscapes, which occupy a substantially smaller part of the region, covering the low-lying areas and lake and river valleys.

The forest-steppe landscapes of the region around the Mayak Production Association are distinguished by the great diversity of their soil cover. The widely encountered forest (dark gray, gray, and light gray), chernozem (meadow, leached, and solonetz-like), meadow-swamp, solonetz, and solodic soils are differentiated by their soil profiles and geochemical parameters. The physical-chemical conditions in the soils vary from weakly acidic oxidizing (in forest soils) to weakly alkaline and alkaline oxidizing-restorative and restorative (in chernozem and saline soils).

During the more than 50-year operating history of the Mayak Production Association, the natural landscapes in the area where the enterprise is located have been subjected to complex radiation and chemical contamination. Analysis of the available data indicates that most of the contamination occurred in the early stage of Mayak's operations, as a result of the above-mentioned radiation accidents in 1957 and 1968, violations of procedures for the safe management of liquid radioactive wastes, and uncontrolled discharges of radioactive wastes into the atmosphere and the lake and river network. The current regulated atmospheric emissions have practically no effect on the environmental radiation situation. Meanwhile, continuing planned discharges of low-level liquid wastes into the reservoirs of the Techa Cascade and medium-level wastes into Lake Karachai are progressively contaminating the rock and groundwater beneath the reservoirs.

Thus, radiation contamination of the soil cover in the natural landscapes and of the bottom sediments in the reservoir-accumulators for liquid radioactive wastes occurred during the initial stage of activities at Mayak's enterprises. The areas affected by radionuclide contamination of the soil may be categorized as follows:

- The East Urals Radioactive Trace, which was created as a result of accidental emissions of radionuclides, occupies a broad territory in the central and northwestern parts of the region being studied. The area's boundary is defined at the 1 Ci/km^2 level on a contour map of strontium-90 activity. In the center of the area, strontium-90 contamination reaches the hundreds of curies per square kilometer.
- Solid radioactive waste storage sites on the Mayak grounds
- Radionuclide anomalies in the Techa River watershed (the Asanov Swamps)

Fairly limited tendencies have been identified for the spread of previously accumulated radionuclides under the physical-chemical conditions prevalent in the forest-steppe landscapes in the Mayak region. Horizontal migration has been noted in the soils of the autonomous landscapes, but since the appearance of radioactive anomalies, this migration has not led to a noticeable change in their initial contours. Vertical migration has been more clearly manifested in these landscapes; nevertheless, most of the radionuclides in vertical soil profiles are attached at geochemical barriers (of the sorption, leaching, or organomineral types) at a depth of 15-20 cm. In superaqueous landscapes, the radionuclides are fixed at deeper levels (up to 50 cm). In contaminated superaqueous landscapes in river valleys affected by seasonal flooding, lateral migration of radionuclides is also noted. This is due to the contaminants being washed out of the soils by floodwaters and carried into the transitory river systems.

The Priargunsk Mining and Chemical Production Association. The deposits of the Streltsov molybdenum-uranium ore field with its extensive and unique uranium reserves are located in the dry steppe zone in the southeastern part of the Lake Baikal region. The PMCPA, which processes these deposits, is Russia's only enterprise that mines and processes molybdenum-uranium ores. The environmental pollutants here include natural radionuclides (uranium-238, uranium-235, and their decay products radium-226 and radon-222), thorium-232, potassium-40, and stable elements (molybdenum, manganese, arsenic, and beryllium), as well as sulfuric, nitric, and other acids.

The area around the enterprise is divided into two major landscape-geochemical systems—the Argunsk (dry mountain steppe) and the Urulyungui (broad intermountain valleys). The terrain in the Argunsk area represents both a source and a transit pathway for flows of substances. They are characterized by a near-neutral, weakly alkaline geochemical environment (pH > 6) and weakly evident vertical and lateral geochemical barriers.

The Urulyungui zone is an area where substances coming down from the mountain massifs accumulate, with some passing through in transit and some being accumulated. Substantial differentiation is noted in the geochemical conditions for migration in both the vertical and the lateral directions, leading to the formation of a wide spectrum of highly absorbent geochemical barriers (restorative hydrogen sulfide and gley types, as well as evaporative and biogeochemical).

The mountainous part of the area is the site of numerous mining industry facilities that are sources of environmental contamination in varying degrees. From an environmental standpoint, the most hazardous among them is the system of tailings repositories at the ore processing complex, which is located in a mountain valley in the northeastern part of the ore field. The repositories contain 30 million metric tons of accumulated solid wastes and 12 million cubic meters of liquid phase wastes, with both varieties containing uranium and its decay products as well as a number of toxic chemical substances. These contaminants move into the surrounding landscape as they are carried by the wind from the repository surface and as liquid wastes components leak through the bottom of their holding vessels.

Anomalous multielement concentrations of the contaminants have been found in the landscapes surrounding the tailings repositories. In contrast, in cases where pollutants have found their way into groundwater, they have formed plumes of contamination affecting significantly larger areas. The main pathway by which contaminants are transferred by groundwater runs away from the tailings repositories and toward the Sukhoi Urulyungui Valley. Therefore, there is a real threat of radiation and chemical contamination of groundwater in the Urulyungui Valley, the source of drinking water for the city of Krasnokamensk. The developing situation requires that the appropriate environmental protection measures be taken.

CONCLUSION

In carrying out radioecological studies in regions where Russia's major nuclear and radiation facilities are located, IGEM specialists for the first time used a new landscape-geochemical method for environmental research developed at IGEM. The studies conducted using this method in the zones affected by five major nuclear and radiation enterprises produced the following results in the first stage of research:

- Detailed studies were made of the morphological, soil-geochemical, and physical-chemical characteristics of the natural landscapes in the areas where the enterprises are located, including in the tundra, forest, forest-steppe, and steppe landscape zones of Russia. Elementary geochemical landscapes were also identified and characterized in each region.

- The primary types of lateral and vertical migration of natural and technogenic radionuclides were identified and studied in each of the types of geochemical landscapes studied.
- The main natural and technogenic geochemical barriers hindering the spread of technogenic contamination were determined.
- An evaluation was made of the degree to which radioactive and stable contaminants arising in the course of operations by nuclear and radiation enterprises affect the environment in the areas where these enterprises are located.
- Methods were developed for small-, medium-, and large-scale radioecological mapping applicable to various landscape-geochemical and geological-geomorphological situations.

By analyzing the results obtained, it is possible to draw the general conclusion that the radiation ecology situation in the regions where the Nerpa Ship Repair Plant and the Smolensk and Balakovo nuclear power plants operate meets current safety requirements. Meanwhile, in the regions where the PMCPA and the Mayak Production Association are located, there are areas where the natural landscapes have been subjected to radiation and chemical contamination. In the area around the Priargunsk facility, the greatest radioecological danger is presented by the repository for the liquid and solid wastes produced by the operations of the molybdenum-uranium ore processing plant and the sulfuric acid production plant. The toxic substances included in these wastes have contaminated the soils, subsoil rock, surface water, and groundwater in the zone affected by the repository. Migration of the water threatens the radiation and chemical contamination of underground sources from which drinking water is drawn.

The wide-ranging plumes of intensive radioactive contamination of the soil cover in the Mayak area (the East Urals Radioactive Trace and the Techa River Valley) were formed in the early stage of activities at the enterprise. During the existence of the East Urals Trace, the contours of this radioactive anomaly have undergone substantial changes. At the same time, lateral migration of radionuclides has been observed in the Techa River Valley caused by the washing out of contaminants from the soil and bottom sediments by flood waters, which have carried the toxic substances into the transitory river systems.

8

Systems Studies of the Radiation Legacy and the Development of the Informational, Legal, and Regulatory Framework for Post-Rehabilitation Institutional Control, Oversight, and Management of Radiation-Hazard Facilities in the Russian Federation*

S. N. Brykin, All-Russian Scientific Research Institute of Chemical Technology
O. G. Lebedev, Russian Research Center—Kurchatov Institute
V. K. Popov, Russian Research Center—Kurchatov Institute
D. A. Serezhnikov, All-Russian Scientific Research Institute of Chemical Technology

As a result of military and civilian nuclear programs during the past 60 years, several countries, particularly the former Soviet Union (or USSR), have created a number of specific engineered facilities and sites that are powerful radiation sources. Large volumes of radioactive wastes have accumulated, and significant areas have been subjected to radioactive contamination. With the end of the Cold War, it became possible to publish substantial information and to begin work on evaluating the contamination and cleaning up affected areas. The idea of studying the radiation legacy of the former USSR was first formulated by the Russian Academy of Sciences (RAS) in 1992. Somewhat later, in March 1993, the government published a book entitled *Facts and Problems Related to the Burial of Radioactive Wastes in Seas Bordering the Territory of the Russian Federation.* This book laid the foundation for an entire range of publications on problems of the radiation legacy.

The first systems study of problems of the radiation legacy of the former USSR was undertaken by the International Institute for Applied Systems Analysis (IIASA). Encouraged by the Russian National Organization for IIASA, the institute created the Radiation Safety of the Biosphere (RAD) project in 1993.

*Translated from the Russian by Kelly Robbins.

One objective of the RAD project was to evaluate the degree and seriousness of radioactive contamination in the former USSR.

In 1995 the International Science and Technology Center (ISTC) approved the start of the RADLEG project (No. 245), Development of a Sophisticated Computer-Based Data System for Evaluation of the Radiation Legacy of the Former USSR and Setting Priorities on Remediation and Prevention Policy. The project was initiated by the RAS, the Russian Ministry of Atomic Energy (Minatom, now the Russian Federal Atomic Energy Agency, or Rosatom), the Russian Research Center—Kurchatov Institute, and IIASA (Austria) and financed by the European Union and Sweden. Completed over the course of 6 years, the RADLEG project aimed to establish a system of data on radiation sources, radioactive wastes, and contaminated areas in the former USSR to facilitate creation of technically and economically effective technologies for cleaning up radioactive contamination. From 1995 through 2001, 24 Russian organizations and agencies, 5 foreign collaborators, and more than 250 individual participants worked on the project.

Based on a simple operational database created in the first phase of the project, a generally accessible RADLEG database was developed in Access and Oracle formats. During the second phase of the project, additional information obtained from both new publications and accessible archives of the participating organizations was added to the database and subjected to expert review. The updated concept for the unified structure of the database includes the following research fields (sectors) involved in the radiation legacy of the former USSR:

- Nuclear power plants
- Shore-based waste repositories, enterprises servicing nuclear power facilities, sunken and submerged objects
- Scientific research institutes, pilot plants, research nuclear reactors, and nuclear research centers
- Nuclear explosions for nuclear weapons testing purposes
- Nuclear explosions for civilian purposes
- Storage and reprocessing of nonreactor radioactive wastes and spent ionizing radiation sources
- Prospecting, mining, enrichment, and reprocessing of uranium ores
- Hexafluoride production and isotopic enrichment of uranium
- Nuclear fuel manufacturing
- Radiochemical reprocessing of spent nuclear fuel
- Production of nuclear materials
- Major radiation accidents
- Power-producing reactor facilities

For each research field, there are screen interface forms with data on the enterprises, organizations, or sites (more than 120 total), each with its own unique code assigned in accordance with an enterprise classification system specially developed under the project. The facility or site is the primary unit in the database structure. Each entry in the database contains a description of a specific facility or site. There are nine main facility or site classifications, each divided into subclasses, the number of which depends on the functional load of the particular facility or site. Each class and subclass is assigned its own unique code. Table 8-1 is an example of the classification system.

TABLE 8-1 Example of the Facility and Site Classification System in the RADLEG Accessible Database

0100	Test sites for underground leaching experiments
0200	Near-surface underground radioactive waste repositories
0300	Deep burial sites for the storage of liquid radioactive wastes
0400	Nuclear reactors and critical stands
0500	Nuclear explosions
0600	Atmospheric emissions
0700	Discharges into water systems
0800	Contaminated lands
0900	Submerged radioactive objects

In addition to the categories "Enterprises" and "Facilities," the user interface also includes "Events" and "Accidents." The "Events" category includes information on the submerging of vessels or other objects, nuclear explosions, contaminated lands, and emissions and discharges of radioactive substances. The "Accidents" category includes data on emergency situations and their consequences. Provisions have been made for the possibility of modifying the database structure and including additional information fields.

An analytical overview entitled "The Radiation Legacy of the Former USSR: Available Data on Main Areas of Research" was prepared in 1996. The paper included information on accumulated radioactive wastes and planned measures for managing radioactive materials. The material was prepared on the basis of data submitted by more than 20 organizations involved in the project. In subsequent discussions, requirements for primary source materials characterizing the situation in each sector of the former Soviet nuclear complex were reviewed and more clearly formulated. It was recommended that separate chapters be prepared on each of the following 11 sectors:

1. Uranium mining and uranium ore enrichment
2. Uranium conversion and isotopic enrichment
3. Nuclear fuel manufacturing
4. Nuclear power plants
5. Nuclear-powered ship engines

6. Research nuclear reactors and nuclear research centers
7. Plutonium production and radiochemical reprocessing of spent nuclear fuel
8. Nuclear weapons production and decommissioning
9. Nuclear explosions
10. Storage and reprocessing of nonreactor radioactive wastes and spent ionizing radiation sources
11. Social aspects of the nuclear legacy

The new edition of the analytical overview devotes special attention to considering the data on the situation regarding the main components of the nuclear fuel cycle. This new edition may be viewed as a fundamental work that has been officially approved by the leading Russian scientific research organizations, in particular, by organizations operating in the nuclear sector. It contains both general information on the nuclear legacy of the former USSR and information on its specific sectors and individual facilities and sites, especially those presenting the greatest radiation hazard to the public and the environment. The review was published by IIASA in February 2000 as a book entitled *The Radiation Legacy of the Soviet Nuclear Complex: An Analytical Overview*.[1]

Organized by the International Atomic Energy Agency (IAEA), the European Commission, the Russian Ministry of Atomic Energy, and the RAS, the international conference "The Radiation Legacy of the Twentieth Century: Environmental Restoration" was held in Moscow in 2000. The conference proceedings (featuring papers by scientists from more than 20 countries, including 15 participants of the RADLEG project) were published by IAEA in 2002.[2]

Further research on the radiation legacy continued under ISTC Project 2097, Development of a Sophisticated Information System Including a Meta-Database and Regional Radioecological Cadastres for Assessment of the Radiation Impact on the Environment and Population: Evaluation Study of the Northwest of Russia and the Krasnoyarsk Region (RADINFO). Financial support for the project was provided by the European Union, the United States, and Rosatom. The results of the RADINFO project (*www.radinfo.org.ru*) included the creation of a meta-database on radiation-hazard facilities and sites in the former USSR; regional geographic information cadastres on radiation sources and radiation contamination in two regions, northwestern Russia and Krasnoyarsk Krai; local geographic information systems (on civilian nuclear explosions, radioecological impacts of nuclear power plants, and so forth); and a demonstration pilot subproject to assess

---

[1]Egorov, Nikolai N., Vladimir M. Novikov, Frank L. Parker, and Victor K. Popov, editors. 2000. The Radiation Legacy of the Soviet Nuclear Complex: An Analytical Overview. London: Earthscan Publications, Ltd.

[2]Radiation Legacy of the Twentieth Century: Environmental Restoration. Papers and Discussions: Proceedings of the International Conference (RADLEG-2000) Held in Moscow, Russian Federation, October 30-November 2, 2000, IAEA-TECDOC-1280. Vienna: IAEA.

the effect of the Mining-Chemical Complex on public health and the environment in the Yenisei River basin, which produced the analytical review "Radioecological Impact of the Mining-Chemical Complex in Krasnoyarsk Krai."

The series of systems studies on the radiation legacy of the USSR under ISTC projects 245 (RADLEG) and 2097 (RADINFO) helped create the necessary informational infrastructure for comprehensively analyzing and evaluating the status of radiation-hazard facilities and sites and for obtaining a graphic picture (not only statistical but also dynamic) of the conditions under which radiation sources operate and the possible means by which radioactive contamination might spread. The research also highlighted the necessary preconditions for assessing radiation impacts on public health and the environment, which in turn opens up prospects for a shift to the next and final stage in studying the radiation legacy. This stage must include development of recommendations for setting priorities in environmental protection policy, including the planning of countermeasures and the formation of a system for monitoring, stewardship, and institutional control over radiation-contaminated areas, which is the ultimate goal of studying the radiation legacy. Plans call for these studies to be carried out under the new RADINKO project (ISTC No. 3707), Development of the Information, Legal, and Regulatory Basis for a System of the Institutional Control and Stewardship of Radioactively Contaminated Areas and Radionuclide Ionizing Radiation Sources in the Russian Federation.

Using the radioecological information system created under RADLEG and RADINFO, an information base for institutional control will be developed. It will consist of local and Internet-based versions of the geographic information cadastre of contaminated areas (using the same data structure established in the worldwide Directory of Radioactively Contaminated Sites produced by IAEA).

The total area contaminated by radionuclides as a result of the activities of the former Russian Ministry of Atomic Energy from 1945 through 2000 is about 480 km^2. A large portion of the contaminated area lies within the boundaries of sanitary-protective zones and observation zones around enterprises. About 15 percent of the total area of radionuclide-contaminated land at Minatom enterprises is characterized by rather high exposure dose rate levels exceeding 200 μR per hour (see Table 8-2). The data presented in the table are from 2000, but these levels have remained practically unchanged to 2007.

Significant areas in the Russian Federation were subjected to radioactive contamination from the Chernobyl nuclear power plant accident in April 1986 and the associated release of a large quantity of radionuclides. Despite the unavoidably subjective nature of efforts to establish demarcation lines in the region to define which areas should be considered radiation-contaminated territory, under accepted practice this area's boundary has been set as the isoline within which initial cesium-137 contamination density is on the order of 37 kBq/m^2 (1 Ci/km^2). In Russia, about 28,000 km^2 falls into this category.

Some areas were contaminated with radiation by nuclear weapons testing at

TABLE 8-2 Radionuclide-Contaminated Areas at Rosatom Enterprises as of January 1, 2000 (in km^2)

Enterprise	Total Area	On Facility Grounds	In Sanitary-Protective Zone	In Observation Zone
Priargunsk Mining-Chemical Association	8.53	7.33	0.78	0.42
Almaz Mining-Metallurgical Enterprise	1.34 (1.03)	1.07 (1.018)	0.27 (0.012)	—
Machine-building plant (Elektrostal)	0.26 (0.26)	0.01 (0.01)	0.13 (0.13)	0.12 (0.12)
Novosibirsk Chemical Concentrates Plant	0.15 (0.14)	0.07 (0.07)	0.08 (0.08)	—
Polymetals Plant (Moscow)	0.016 (0.001)	0.002 (0.001)	0.014	—
Chepetsk Mechanical Plant (Glazov)	1.35 (0.062)	1.34 (0.059)	0.01 (0.003)	—
Zabaikalsk Mining-Enrichment Complex	0.04	0.04	—	—
Mayak Production Association	452.16 (65.70)	38.46 (17.70)	217.54 (38)	196.16 (10)
Mining-Chemical Complex (Zheleznogorsk)	4.71 (0.203)	4.29 (0.19)	0.07 (0.013)	0.35
Siberian Chemical Complex (Seversk)	10.39 (4.191)	10.09 (4.026)	0.30 (0.165)	—
Kirovo-Chepetsk Chemical Complex	0.70	0.17	0.15	0.38
All-Russian Scientific Research Institute of Technical Physics (Snezhinsk)	0.13 (0.01)	0.13 (0.01)	—	—
Scientific Research Institute of Atomic Reactors (Dmitrovgrad)	0.39 (0.081)	0.15	0.24 (0.081)	—
Institute of Physics and Power Engineering (Obninsk)	0.001 (0.001)	0.001 (0.001)	—	—
Total	**480.32** (71.68)	**63.235** (23.08)	**219.64** (38.48)	**197.43** (10.12)
Excluding Mayak	**28.16** (5.98)	**24.79** (5.38)	**2.10** (0.48)	**1.27** (0.12)

NOTE: Figures in parentheses represent the area where the exposure dose rate exceeds 200 μR per hour.

the Novaya Zemlya test site or by nuclear explosions carried out in other locations for industrial purposes (particularly in the Republic of Yakutia and Ivanovo and Perm oblasts). All told, 81 underground nuclear explosions were set off for economic purposes in the Russian Federation from 1965 through 1988 as part of the program Nuclear Explosions for the Economy. All of the sites in the Russian Federation that were created using nuclear explosion technologies may be placed in one of the following four categories:

1. Sites closed by appropriate legal acts: 40, including 2 (Dnepr-1 and Dnepr-2) closed under officially approved projects
2. Sites in operation for their original purposes: 16, 2 of which (Kama-2 and Vega, vessels 1T-15T) have operating organization status
3. Sites at the decommissioning stage: 21, 15 of which are being studied
4. Sites in catastrophic status (Globus-1 and Kraton-3) or categorized as such (Taiga and Kristall): 4

Undoubtedly, the greatest danger to the public living near sites created using nuclear explosion technology is that resulting from the catastrophic civilian nuclear explosions Globus-1 and Kraton-3, as well as the explosions involving soil dispersal (Taiga and Kristall), which specialists also classify as catastrophic. It should be noted that localized radioactive contamination is found on the grounds of about 20 sites in all four categories, and there are also cases of burial of soil, mine tailings, and equipment. As of 2007, new problems and even contradictions have been uncovered in connection with the institution of new laws in the Russian Federation covering such areas as nuclear energy, public radiation safety, environmental protection, and others, and as a result of new conclusions drawn from data on the radiation consequences of nuclear explosions.

The former onshore technical bases serving the nuclear naval fleet on the Kola Peninsula (Andreev Bay, Gremikha village) represent real sources of radiation risk to the public and the environment. Furthermore, Russia has areas subjected to radioactive contamination from prospecting, mining, and processing of uranium and thorium ores, which have led to increased concentrations of natural radioactive substances. The urgent task today is to complete a full inventory of such areas as part of the unified federal system for accounting and control of radiation-contaminated areas. This will create the necessary preconditions for adequately monitoring the radiation situation in these areas and adjacent zones and developing rehabilitation programs.

Russia has now accumulated a certain amount of practical experience in rehabilitating radiation-contaminated areas. This includes both areas subjected to contamination from the Chernobyl accident and lands contaminated by the operations at such sites (including incidents involving the release of radioactive substances into the environment) of nuclear industry facilities like the Mayak Production Association in Chelyabinsk Oblast, the Siberian Chemical Complex in Tomsk Oblast, and the Almaz Uranium Mining Enterprise in Stavropol Krai. Experience gained in rehabilitating areas at the Russian Research Center—Kurchatov Institute, located in the midst of a residential area in northwestern Moscow, and at the Moscow Polymetals Plant merits special attention. In many cases, rehabilitating sites to the extent that radionuclide concentrations are reduced to natural background levels, thus allowing them to be used afterwards without restriction or constant monitoring, is impossible for several reasons (technical limitations, economic expedience, worker health considerations, safety problems, prevention of environmental side effects, and so forth). This means that institutional control must be maintained over such sites for a prolonged period.

Long-term post-rehabilitation oversight entails a differential approach to organizing systems of measures to manage and monitor sites after restoration work is completed, including establishing administrative responsibility, a legal regime, environmental monitoring systems, and safety measures. The legal framework for this sort of activity in the Russian Federation has yet to be clearly defined; therefore, a top priority task is to analyze the Russian Federation's existing legal, regulatory, and methodological documents related to the rehabilitation of radionuclide-contaminated areas and the subsequent monitoring of their status. An optimal balance must be found between social and economic factors on the one hand and the degree of protection on the other.

In recent years the IAEA has initiated a broad program of efforts touching on all aspects of environmental rehabilitation. In particular, drawing from the experience of the United States, Great Britain, Germany, Canada, and other countries in this field, the agency is conducting analyses and holding international meetings under the heading "Management of Long-Term Radiological Liabilities: Stewardship Challenges." Part of this activity involves developing a new approach to the problem of reducing the radiation risk from the effects of sites left after the decommissioning of radiation-hazard facilities. This approach entails shifting away from the idea of rehabilitating the site to "green field" status and instead establishing long-term institutional monitoring.

Based on the regulatory and legal framework, the creation of which was also a task under the project, categorization rules and criteria will be developed as well as classification principles (by levels of potential danger to the public and the environment) for areas subjected to radioactive contamination as a result of the activities of Rosatom enterprises. Data will be systematically organized on the regulatory, legal, and informational content for a system of accounting and institutional control over ionizing radiation sources in the Russian Federation. Plans also call for the creation of a project Web site with a system for assigning user access rights to the information contained. An analytical review will be prepared of Russian experience in the environmental rehabilitation of radiation-contaminated areas, including the organizational, legal, technological, medical-sanitary, economic, and social aspects of such efforts. Based on the information infrastructure and on an analysis of existing domestic and foreign regulatory and legal documentation, the foundations will be laid for a computerized decision support system regarding forms and limits of institutional control. The final stage in the project will be the development of proposals on the formation of a strategic plan for long-term institutional control (including after rehabilitation) over radiation-contaminated areas in the Russian Federation (taking IAEA developments into account).

9

Comprehensive Resolution of the Problem of Radioactive Waste Management and Rehabilitation of Contaminated Areas in the Moscow Region*

S. A. Dmitriev, Moscow Joint Environmental-Technological Scientific Research Center for Radioactive Waste Decontamination and Environmental Protection (MosNPO Radon)

MEASURES TO IMPROVE THE RADIOACTIVE WASTE MANAGEMENT SYSTEM IN RUSSIA

Analysis of possible means of achieving the objectives formulated in the document *Principles of State Policy on Ensuring Nuclear and Radiation Safety in the Russian Federation through 2010 and Beyond*, approved by the President of the Russian Federation, highlights several very significant problems and shortcomings associated with state policy on radioactive waste management, and the development and security of the country demand that they be resolved.

The complexity and scope of the tasks associated with radioactive waste management are determined by the following 10 circumstances:

1. Despite certain actions that have been taken to develop and implement specific programs for reprocessing and utilizing radioactive wastes, the country's unified policy on radioactive waste management remains in a formative stage.
2. There is still a prevailing tendency to put off fundamental decisions and actions on long-term state programs that have been developed regarding scientific

*Translated from the Russian by Kelly Robbins.

research, creation of a modern regulatory base and new technologies, and development of the technical infrastructure for waste management and isolation.

3. An increasing number of radiation-hazard facilities in the country need to be removed from operation (nuclear-powered naval and civilian ships, nuclear reactors at atomic power plants built in the 1960s, research and training reactors, irradiation equipment, monitoring devices that use radiation, and other types of radiation-hazard equipment).

4. The city of Moscow and the Central Region of the Russian Federation require significant attention. Many radioactive waste-producing facilities are concentrated there (representing 70 percent of all wastes from nonnuclear applications produced in Russia), and at the same time it is an area of maximum population density.

5. Land rehabilitation and reclamation is necessary in areas where uranium ore is mined and mine tailings are currently stored, and petroleum drilling enterprise wastes with high concentrations of natural radionuclides must be reprocessed.

6. Work is required to rehabilitate areas where the radiation situation is unfavorable, formed as a result of radiation accidents. Monitoring of the radiation and environmental status of cities and smaller towns is also necessary, as well as forecasting and prevention of radiation-related extreme events.

7. The current situation is characterized by the presence of the negative so-called nuclear heritage, including such problems as the accumulation of wastes in the Techa Cascade and Lake Karachai at the Mayak Production Association, and so forth.

8. Existing regulatory documents on radioactive waste management need improvement. There is a lack of experience in standard solutions and technologies with a high level of reliability and safety.

9. Work on a nationwide scale to create a unified waste isolation system began in the 1970s-1980s, but lags significantly behind world practice. Indeed, the greatest successes in radioactive waste management have been achieved in countries that have a unified waste management system. These states have specially created governmental organizations: ANDRA in France, ENRESA in Spain, NIREX in Great Britain, and so forth. They operate independently of the radioactive waste producers, are directly subordinate to the country's government, and establish requirements for quality and minimization of radioactive wastes. National radioactive waste repositories with capacities up to hundreds of thousands of cubic meters operate under their auspices. However, foreign experience of countries with unified radioactive waste management systems (Great Britain, France, Spain, Belgium, the Netherlands, and others) is being insufficiently studied and ineffectively used.

10. There is inadequate scientific, organizational, and technological coordination of efforts in radioactive waste management. Enterprises that produce

such wastes may be found in such industries as energy, manufacturing, defense, mining, medicine, agriculture, space, and science.

REHABILITATION OF LAND AREAS AND DECOMMISSIONING OF RADIATION-HAZARD FACILITIES (THE EXPERIENCE OF MOSNPO RADON)

In terms of its structure and specialization, the Moscow Joint Environmental-Technological Scientific Research Center for Radioactive Waste Decontamination and Environmental Protection (MosNPO Radon) is the closest in the Russian Federation to the aforementioned types of efforts involved in comprehensively addressing radioactive waste management problems. The enterprise was created as part of a special project, and during the almost half a century since its founding, it has developed as a necessary and important part of the city's structure for ensuring its technogenic,[1] environmental, informational, and social security and for addressing questions related to centralizing radioactive wastes from organizations in various industries across a significant part of the Russian Federation.

The enterprise's significance to Moscow and the Central Region is based on the capital being home to more than 20 nuclear- and radiation-hazard enterprises and more than 600 organizations that use ionizing radiation sources in their operations. There are 37 nuclear-hazard facilities of various types within the city limits. The operations of these organizations produce a significant quantity of wastes. Radiation-contaminated areas, a legacy from past decades, are being discovered in the city, and they require constant radiation monitoring. To address this problem, MosNPO Radon has developed and instituted a system of radiation-ecological monitoring (with about 200 sampling points) that ensures the necessary radiation safety of the metropolis. Relevant information is also conveyed to the public, creating a favorable moral and psychological climate. Unique technological processes and equipment have been created at the enterprise, and the Moscow radiation monitoring system is the only one in the Russian Federation that covers the territory of an entire major city.

MosNPO Radon is currently a multipurpose, multifunctional complex, and its "radioactive waste storage point" is in fact one of the numerous systematic elements within its structure that comprehensively ensure the radiation security of Moscow and the country's Central Region. A modern scientific research infrastructure has been created, and the enterprise conducts research and experimental design work aimed at improving the safety of radioactive waste management. It has also amassed experience in the development of regulatory documentation and radiation-monitoring systems for use in the technological processes involved in radioactive waste reprocessing.

[1] *Technogenic* is used to refer to phenomena arising as a result of the development or deployment of technology.

The Center for Radioactive Waste Reception and Transport and Radiation Emergency Response was organized in 1999. The center carries out the following work in the decontamination and decommissioning of radiation-hazard facilities: development of decontamination plans, regulations, methods, and methodologies; decontamination of affected areas and facilities; and shipment preparation and transportation of radioactive wastes. The center's main tasks include the following:

- Removing radioactive wastes from enterprises in Moscow and Russia's Central Region
- Deactivating radiation-contaminated sites and deactivating and decommissioning contaminated facilities
- Developing regulatory, methodological, and technical documentation
- Transporting radioactive wastes
- Conducting searches for radioactive sites, radiation monitoring, and other radiation-related studies
- Decontaminating individual protective devices and specialized clothing
- Developing new methods for deactivating and dismantling various types of structural elements

Since its establishment, the center has carried out the following work:

- Deactivated more than 1,000 radiation-contaminated sites
- Removed more than 23,000 m^3 of radioactive wastes on orders from enterprises in Moscow and Moscow Oblast as well as other facilities in Russia's Central Region
- Handled more than 950 emergencies involving radiation accidents
- Removed more than 450 ionizing radiation sources
- Decommissioned more than 10 contaminated facilities

Moscow Polymetals Plant—Open Joint-Stock Company

A radiation and environmental study was carried out at 32 buildings and structures and adjacent grounds and production space at this industrial enterprise in Moscow, which covers about 165,000 m^2. The study uncovered more than 400 radiation hot spots at the site, with a total area of 9,150 m^2. The gamma radiation exposure dose rate at the surface of the radiation hot spots reached 4,000 μR per hour with an alpha-particle flow density up to 19 particles per square centimeter per hour. Thorium-232 and radium-226 were the primary radionuclide contaminants.

From 1999 through 2002, 17 contaminated buildings and practically the entire contaminated grounds at the enterprise were deactivated. Buildings in which work had been done with radioactive substances were also dismantled. During the deactivation process, the radioactive wastes were prepared for transport into long-term storage, a process that included fragmentation of large pieces, conditioning, and packing of wastes into metal or plastic containers. Some 425 m^3 of radioactive wastes were prepared and shipped off for long-term storage.

Deactivation work at radiation-contaminated sites proceeded to a depth of up to 1 m by means of manual sorting of extracted soil. Excavated areas were filled in with clean soil or paved with concrete topped with asphalt. As a result of the decontamination of these sites, the gamma radiation exposure dose rate was reduced to background levels (11-18 μR per hour). Decontamination of the buildings was highly effective, also reducing the gamma radiation exposure dose rate to background levels, an average of 15 μR per hour (0.15 μSv per hour), reaching maximum levels of 27 μR per hour (0.27 μSv per hour) in certain points. Following completion of the rehabilitation efforts, the individual effective dose from external exposure for plant workers will not exceed 0.3 mSv annually.

Carrying out this decontamination work has made it possible to develop a methodology, organizational approaches, and forms of cooperation among various organizations involved in radiation accident response efforts at enterprises in the nuclear complex. In the course of the work, new deactivation technologies have been improved and developed, and much experience has been gained in addressing the problem of rehabilitating a major contaminated Russian Federal Atomic Energy Agency (Rosatom) facility.

The Radiochemical Laboratory of the Vernadsky Institute of Geochemistry and Analytical Chemistry

This laboratory operated in Moscow from 1966 through the end of the 1980s. It included 7 hot rooms and associated control rooms with an area up to 240 m^2, 15 auxiliary rooms, a ventilation system, and so forth. The project had to take into account that the laboratory building had to be rehabilitated without changing the structure of the first floor and without dismantling the technical equipment, special structures, and utility lines. The decontamination work proceeded at the same time that the building was being renovated. While work was under way, an analysis of the situation indicated that the requirement to dismantle the equipment contradicted the demand to preserve the building's structure, which led to the need to carry out additional deactivation cycles and to apply treatments to preserve the building structure for 50 years. Experience was acquired in working in radiochemical laboratory facilities with high levels of contamination. As a result of the intensive deactivation effort and in accordance with the requirements of the client, maximum dose levels were reduced to 20 μR per hour as opposed to maximum initial readings of up to 68,000 μR per hour.

Open Joint-Stock Company Kolchugtsvetmet [S. Ordzhonikidze Kolchugino Plant for Ferrous Metals Processing]

Located in Vladimir Oblast, the Kolchugtsvetmet plant was contaminated with natural radionuclides (radium-226). The contaminated area included a two-story, 1,200 m^2 production building. The first floor measured 898 m^2 and the second, 317 m^2. The production building was constructed and put into service in the early 1950s to manufacture phosphorescent substances using a radium-226 bromide solution. Radioactive materials were involved in the production process in both solid and solute form and were used in dust separation operations. A radioecological study of the laboratory facilities revealed radiation-contaminated areas of more than 200 m^2, with a gamma-irradiation exposure dose intensity from 45 to 28,000 μR per hour. Removal of weakly attached contaminants and treatment of the surfaces of equipment, walls, and floors to reduce the risk of personnel contamination were accomplished using Optimist-type deep-penetrating primer. The walls were then treated with acrylic façade paint. The floors, walls, and equipment were treated in areas with surface-fixed alpha contamination of more than 5 alpha-particles per minute per square centimeter. Some 500 m^3 of radioactive wastes were removed from the facility.

At this site, the center gained experience in working under high-level contamination conditions. A brick building was completely cleaned, and after radiometric monitoring it was dismantled, along with its autonomous ventilation system and pipes. The site where the dismantled building once stood was then rehabilitated along with adjacent areas.

The Kurchatov Institute

Wastes with a total activity of several thousand curies and a volume up to several thousand cubic meters are present on the grounds of this institute or could be created there as a result of decommissioning efforts, and this situation requires significant time for waste reprocessing. MosNPO Radon is carrying out this work rather successfully. Given the urgency of the problem of reducing the volume of radioactive wastes subject to long-term storage, a technology for reagent-based decontamination of soil and related materials has been developed from research and lab tests. This technology will make it possible to reduce by dozens of times the volume of radioactive soil that must be stored. It includes a complete cycle for leaching radionuclides from the soil at increased temperature in an agitation regime using a mixed reagent solution with subsequent reprocessing of the resulting technological solutions. The soil is thus cleaned to meet regulatory standards and can be removed from the enterprise's production zone. The radioactive concentrate that formed as a result of the cleaning process is sent away for long-term storage.

Specially created units with simultaneous processing capacities of 12, 60, and

1,000 kg of contaminated soil were used to work out this technology for reagent-based soil decontamination in a rotary reactor-mixer and to test its capabilities in various operating regimes and output capacities. While these tests were being conducted, work was simultaneously under way to develop means of decontaminating technological solutions and neutralizing water used in washing operations. The tests were conducted on real batches of cesium-137–contaminated soil taken from the grounds of the Kurchatov Institute. The following values for the basic technological parameters involved in reagent-based soil decontamination were determined during the tests: the composition and concentration of the sulphuric acid-based reagent solution; the process temperature, 95°C; and the liquid-solid phase coefficient, L:S = (1:1.5)/1. Under these parameters, 95-98 percent of the contaminants are removed from the soil.

Decontamination of the technological solutions is achieved by means of ferrocyanide precipitation of cesium. After adjustment of its composition, the cleansed solution has a sufficiently low specific activity level and may be reused for processing the next batch of soil. Organizing the soil deactivation process so as to utilize the technological solution in a closed cycle makes it possible to reduce reagent requirements significantly to 50 kg per metric ton of soil.

As a result of lab-based and field testing, the technological process for reagent-based removal of cesium-137 from soil was developed. The main apparatus used is a rotary electric-heated reactor-mixer used for processing the soil with reagents. This equipment is currently being tested for use in removing radionuclides from soil. The tests involve checking the technical features included in the design of the apparatus and the operating capacity of this nonstandard equipment. Various deactivation procedures are also being worked out.

Two batches of soil actually contaminated with cesium-137 have been processed. Processing of the first batch of 535 kg of soil reduced the equivalent dose level by 72 times to 10 μR per hour, which corresponds to the background level, while specific activity was reduced by 40 times to 0.51 kBq/kg.

DISCUSSION OF WORK EXPERIENCE

As a result of the activities of MosNPO Radon, practically all points of surface contamination discovered in Moscow have been eliminated. At the same time, substantial characteristic problems require attention. At almost all facilities that were in operation in the 1940s-1950s, there is no reliable information on the structure of the buildings, their condition, and the location and condition of utility lines and pipes. At practically all old enterprises still in operation, there is also no information on the presence of hot spots due to accidental spills of radioactive substances or unauthorized radioactive waste burial sites on their grounds. Therefore, any project must begin with a search for any information that has been maintained and a full-scale radiation study of the facility.

The experience amassed by MosNPO Radon attests to the city having radiation-contaminated sites that have been concealed by new construction, erosion of riverbanks and other slopes, and other construction-related activities. Ensuring radiation safety for the public demands that such hot spots be discovered and eliminated.

In preparing for work aimed at rehabilitating sites in the city of Moscow, it must be kept in mind that ignorance of what awaits construction workers carrying out site preparation work, especially at locations formerly used as industrial dump sites, not only could lead to significant expense but also could be a potential source of radioactive contamination for the surrounding area. Moscow currently has about 2,000 facilities in operation that work with ionizing radiation sources and radioactive substances. However, the city has no comprehensive database about enterprises that previously worked with ionizing radiation sources but closed or changed orientation as a result of economy policy changes during the past 15 years.

From the above-mentioned points, we may conclude that systematic data are needed on a citywide scale regarding the locations of potentially hazardous radiation-contaminated sites. Therefore, it is proposed that a program be developed to conduct a comprehensive radiation study of facilities where work with radioactive substances has been carried out since the late 1980s. Such a program should also include studies of known and suspected sites of soil contamination at specific facilities and other areas of the city.

Case Studies

10

Lands Damaged as a Result of Uranium Ore Mining Operations in the Russian Federation

V. P. Karamushka and V. V. Ostroborodov, All-Russian Research, Design, and Surveying Institute of Industrial Technology (VNIPIPT)

The operations of uranium ore mining and processing enterprises inevitably contaminate the environment with solid, liquid, and gaseous wastes, which are the most voluminous within the nuclear fuel cycle industries and, despite their relatively low activity level, significantly contribute to the radiation hazard facing the population.

Table 10-1 provides data on radioactive contamination of the main sectors of the environment at uranium ore mining enterprises.

Beginning from the time large-scale uranium ore geological prospecting and mining began in the USSR in 1945, nine industrial complexes were built on the territories of six Soviet republics. The complexes now operate as production facilities in the Newly Independent States. The total land area damaged by their production activities is 80,500 ha, and about 220,000 ha may be affected eventually by radioactive products transferred by air streams, surface water flows, and so forth.

At this time in the Russian Federation, both currently operating facilities and worked-out deposits need to be put in safe condition from the standpoint of radiation risk to the population.

Following is a list of decommissioned mining enterprises of the former Russian Ministry of Atomic Energy (Minatom), as well as geological prospecting and experimental production facilities where operations have been discontinued:

TABLE 10-1 Characterization of Radioactive Contamination of the Main Sectors of the Environment

Sector	Radiation Hazard Factors	Unit of Measurement	Background Values		Maximum Permissible Concentrations Above Natural Background	Actual Recorded Value Range	Notes
			Average	Range			
Air	Equivalent equilibrium concentration of radon daughter products	Bq/m^3	1.8	0.1-10.0	37[a]	185-260	At distance of 100-150 m from source
	Total $\acute{\alpha}$-activity of long-lived radionuclides of the uranium natural radioactive decay family	mBq/m^3	0.01	0.001-10	12.0	4-40	At distance of 100-150 m from source
Water	Ra-226 and daughter products	Bq/m^3	37	1.11-111	111[b]	300-500	During geological prospecting and preparatory mining operations
						3,700-5,500	During cleanup operations
						10,000-50,000	In crude mine waters
						10,000-80,000	In liquid phase of hydrometallurgical plant (HMP) tailings
Soils	Total $\acute{\alpha}$-activity of long-lived radionuclides of the uranium natural radioactive decay family	Bq/kg	370	90-700	600-1,200	1,000-7,400	In sanitary protective zone of operating enterprises

Solid Waste							
Refuse heaps	ά-activity	Bq/kg	240	150-10,000	600-1,200	1,000-12,900	Refuse heaps, waste piles, radiometric enrichment and heap-leaching tails
	γ-activity	μR/hr	12	10-60	20	20-200	
HMP tailings	Radon exhalation from surface	$Bq/m^2/s$	0.016	0.004-0.053	1.0^c	1.7-30.0	On dry beaches of tailing dumps
	Solid phase ά-activity	Bq/kg	210	370-700	600-1,200	1,100-27,300	
	Solid phase γ-activity	μR/hr	12	10-60	20	100-1,400	

[a] = outdoors, for dwellings: 200 Bq/m^3
[b] = water reservoirs, for wastewaters assuming their dilution: 185 Bq/m^3
[c] = USSR standard, by ICRP data: 0.74 $Bq/m^2/s$

- Lermontov Industrial Association Almaz
- Novotroitsk Thorium Ore Mining Administration in the city of Balei, Chita Oblast, and an open-cast mine near Ozernoye settlement
- Stepnoye Ore Mining Administration in the Republic of Kalmykia
- Numerous experimental production sites where geological prospecting parties operated, the biggest among them being the Aldan deposit in the Republic of Sakha (Yakutia)

The Lermontov Industrial Association Almaz is most significant from the standpoint of mine reclamation activities. We shall now consider it in more detail, as the situation there demands immediate action. The enterprise is located in the Mineralnye Vody spa area in the Caucasus, close to the city of Lermontov, Stavropol Krai (130 km southeast of the city of Stavropol).

The Almaz site effort is a model pilot project for the development and implementation of a comprehensive program for remediation of land and facilities left behind after uranium ore mining and processing operations were discontinued at similar enterprises in the Russian Federation. The Almaz uranium ore mining and processing enterprise was established in 1950 on the basis of a local mining and chemical ore mining administration. Its underground mines Beshtau, Byk, Shargadin, and Stepnoye were the first to be set in operation. In Lermontov in 1954, a hydrometallurgical plant (HMP) began processing uranium ore with triuranium octaoxide as its final product. The plant also produced fertilizers, scandium oxide, and some other by-products. To remove radioactive waste from the plant, tailings dumps were created a few kilometers away.

Mine No. 1 ceased operation in 1975, and decommissioning work continued to be performed there until 1994. An electromechanical plant was built on the former mine site. In 1989, Mine No. 2 was also put out of operation. Its mining equipment was dismantled in 1991 and reclamation of the mine's galleries was subsequently carried out. In 1991 the HMP was shut down, its production equipment was dismantled, and the main building of its uranium processing plant was decontaminated. The radioactive waste, including contaminated equipment, was removed to the tailings dump.

After all uranium ore mining work ceased and the equipment was dismantled, the natural groundwater level began to be gradually restored. Filtration of significant volumes of groundwater through highly mineralized rocks, which had dried as a result of the ore mining operations, now creates a real hazard as it contaminates the water with radionuclides and other chemical components. There is a rather high probability that surface water (the Kuma River system) and sources of fresh and mineralized groundwater in the Mineralnye Vody spa area in the Caucasus will be contaminated.

The sites of mines No. 1 and 2 and the HMP tailings dump pose direct hazards to the local population and environmental safety. Contamination is caused by uranium, its daughter products, and other nonradioactive substances penetrating into the system of surface water and groundwater, as well as by radon emanation.

At present, the upper sections of the HMP tailings dump are covered with a layer of apatite-processing waste (neutralized phosphogypsum), which reduces the passage of radon. The dump's lower section is almost completely covered with surface water. The results of a preliminary assessment indicate that radon emanation from the HMP tailings dump is rated first on the list of radiation impact factors, with the tailings pile from Mine No. 2 taking second place. The high level of radiation impact is caused by the contamination of the hydrogeologic system as a result of radionuclides and other chemical substances leaching from the uranium ore.

The underground mines are accessible. The total amount of radioactive waste there is about 22 million metric tons with about 49,000 Ci of total activity. Mine conservation and reclamation works have been performed occasionally.

According to radiation-monitoring data, the gamma-radiation background level in the tailings dump area is within the range of 200-2,000 μR per hour (in some places up to 4,000 μR per hour). The average soil alpha-activity level does not exceed 2,400 Bq/kg in the tailings dump area and 800 Bq/kg at the HMP site.

Taking into account the particular importance of the Mineralnye Vody spa area of the Caucasus (in accordance with the Russian president's Decree No. 309 of March 27, 1992, entitled "On a specially secured ecological and resort region of the Russian Federation") and with the goal of effectively applying European experience in uranium waste management in the European Union (EU) and East European states, the Russian government decided to ask the EU to provide technical assistance in developing an adequate program for the former uranium ore mining facilities in the Lermontov area for management and reclamation.

A concept has now been developed for uranium ore mine reclamation at the former state-owned enterprise Almaz in Lermontov. This plan was prepared within the framework of an EU program by a consortium including Wismut GmbH (Chemnitz), Wisutek GmbH (Chemnitz), C&E GmbH (Chemnitz), G.E.O.S. GmbH (Freiburg), Atomredmetzoloto Open Joint-Stock Company (Moscow), VNIPIPT (Moscow), and the Hydrometallurgical Plant (Lermontov). The concept (EUROPAID/116483/C/SV/RU) is the last step preceding creation of an ecologically safe zone on the territory of the former uranium ore mining enterprise Almaz in the Mineralnye Vody region.

The main results attained under this program are as follows:

- A database has been created using up-to-date computer technology.
- The need for carrying out reclamation works has been established and the necessary protection limits have been assigned.
- Remediation options have been determined with raw economic surveys for the following complex objects:

– Beshtau mine
– Byk mine
– Hydrometallurgical plant
– HMP tailings dump

Social aspects of the resort region's development, as well as concerns of the Lermontov city residents and the local administration, were also taken into consideration.

In accordance with the plan, a guaranteed supply of radon water to the resorts of Pyatigorsk is foreseen, and radiation and environmental safety is to be ensured for both the local population and the recreational areas near Lermontov. The next step is practical implementation of the plan, including elaboration of working projects and their realization in concrete form. A project implementation schedule is currently being developed. To create plans for a working project for reclamation of the HMP tailings dump, a significant amount of additional data on the physical and mechanical characteristics of the deposits will be needed.

In all enterprises of the former USSR Ministry of Medium Machine-Building (Minatom was its successor), there were special service units engaged in environmental (radiation inclusive) and geochemical monitoring of water conditions. There was a special research laboratory at VNIPIPT engaged in developing technologies for tailings dumps and monitoring their conditions. As a result of the activities at VNIPIPT, which had planned all of the uranium mining enterprises both in the USSR and in Eastern Europe, the institute created an unprecedented database on such capital-intensive (as far as restoration efforts are concerned) objects as tailings dumps, the world's largest database of its kind.

The research carried out made it possible to develop methods for underground water contamination forecasting. The methods were used in practice and showed a good convergence of the forecasts produced with actual operational monitoring data.

In the period after a facility ceases operation, the main factor determining the reliability of technical solutions for tailings dump reclamation (conservation) is the knowledge of physical and mechanical characteristics of tailings sediments both to date and for a long time to come. Such data are partly available at VNIPIPT, as a result of studies that have been carried out.

Systematic analysis of data from operational observations indicates the need for diversity in dealing with sites located in the arid zone and areas where the climate is continental or extremely continental.

The available database on sites in the Commonwealth of Independent States (CIS) makes it possible to create an atlas of physical and mechanical characteristics of tailing sediment materials and to use simplified testing procedures for the materials before performing reclamation works. Such work is already being carried out for the Central Asian region. The suggested approach makes it possible to avoid rather expensive and dangerous operations at the tailings dumps and to save on expenses for studies by several thousands of dollars.

The next step is the development of procedures for evaluating the status of worked-out tailings dumps for various climate zones and highlands. For that purpose, we have proposed some procedures for evaluating the radioecological safety of tailings dumps as well as a procedure for assessing the environmental efficiency of reclamation works. The development of a package of standard technologies for the CIS countries with necessary amendments to the regulatory documents should be the final stage of the efforts.

As the data from observations made at both operating and worked-out tailings dumps since 1969 show, the particle size distribution of the tailings, ore processing technology, and climate conditions are the determining factors in the process of forming strength properties.

At the sites under consideration, the following uranium extraction processes were used:

- Acidic
- Sodium carbonate treatment
- Acidic leaching from burnable coal zone

As to the effect of climate conditions, it was established that each of the climate zones requires an individual approach, and in each of them, specific reclamation technologies should be used. This can be shown by the example of sites located in arid zones. According to the 1997 United Nations Educational, Scientific, and Cultural Organization classification, arid territories on the earth are subdivided into four bioclimate zones, each characterized by its aridity index value:

1. Extra-arid zone: annual rainfall less than 100 mm
2. Arid zone: annual rainfall within the range of 100-200 mm, aridity index <0.03
3. Semiarid zone: annual rainfall within the range of 200-400 mm, aridity index 0.2-0.5
4. Zone of insufficient moistening: annual rainfall within the range of 400-800 mm, aridity index 0.5-0.75

By that classification, most of the uranium ore mining enterprises of the former USSR are situated in the arid and semiarid zones and in the zone of insufficient moistening. Thus, all such facilities located in Kyrgyzstan, Uzbekistan, Tajikistan, and, partly, Kazakhstan fall into these categories. There are such enterprises in Russia as well: the Stepnoye Ore Mining Administration in the Republic of Kalmykia and the Lermontov Industrial Association Almaz.

For the tailings dumps, the main distinction between the arid zone and the areas with continental and extremely continental climate is linked with humidity conditions in the mass of tailing sediments.

If the climate factor is not taken into account, a water lens can form within the body of the tailings dump. According to studies that have been carried out, the lens formation mechanism to a certain degree depends on desert groundwater recharge conditions through the suspended water zone. As our studies show, a water-saturated zone can form over a very long period in the tailings mass. The zone is isolated from the surface by a mulch layer, preventing evaporative processes.

Available factual data on the dependence of tailing sediment humidity on particle size, density, and stabilization time make it possible to develop a method for predicting humidity conditions over the long term. We are currently engaged in such work. On the basis of results of the forecasting efforts, technical solutions should be taken to ensure that saturated surface egress to the downstream side of the filled-up tailings dam is impossible.

In our practice, one solution to this problem is to make an absorbing well by means of horizontal drilling. The water-saturated zone is discharged through the well. The place for drilling is determined on the basis of forecasting calculations, and the well is monitored by penetration studies.

Results of the planned efforts to create an atlas of tailing sediment charts and geochemical landscapes for the CIS countries may also be used for East European sites as well as for other industries where environmental contamination with natural radionuclides occurs.

Taking into account the large scale of the efforts and their significance, we believe that the work can be an object of international cooperation within the framework of European Commission and International Science and Technology Center programs.

11

Uranium Recovery and Remediation of Uranium Mill Tailings: Russian and U.S. Experience

James H. Clarke and Frank L. Parker, Vanderbilt University

INTRODUCTION

Recovery of uranium using conventional mining—open pit and underground—to excavate the ore and chemical processing to extract the uranium from the ore generates large volumes of solid and liquid material (uranium mill tailings, or UMT). These wastes are typically collected and stored in ponds, although there has been some effort to replace them in the excavated areas. If left uncovered, UMT constituents can be released to the atmosphere and transported to surface waters and groundwater through erosion and leaching, respectively. Remediation-reclamation approaches either keep the tailings covered with water, leave the dewatered tailings in the ponds or piles, or place them in prepared near-surface or surface pits with drainage and covers to reduce the erosion, resuspension, and leaching of the tailings.

Until the mid-1970s, uranium was recovered exclusively through conventional mining approaches (underground mines and open-pit mining) in the United States. The first commercial application of in situ recovery and in situ leach (ISL) techniques that used chemical agents to extract the uranium through an array of injection and recovery wells was in 1975. Use of this technology increased as it gained a cost advantage, and now most uranium in the United States is recovered using ISL. In situ approaches have the advantage of eliminating large volumes of tailings. However, it is typically impossible to restore groundwater quality to its initial state when uranium recovery is done in situ or to totally contain all of the leachate material (Davis and Curtis, 2007). However, most uranium mining wastes in the United States have come from conventional mining operations, both surface and underground, from 1957 to 1989 (World Nuclear Association, 2007).

URANIUM RECOVERY AND MILL TAILINGS IN RUSSIA AND THE FORMER SOVIET UNION

Uranium recovery in Russia and the former Soviet Union is done through both ISL and conventional mining. In the early 1990s, approximately 38 percent of uranium mined in the former Soviet Union came from ISL (Bradley, 1997).

For the Russian Federation, on the basis of metric tons of uranium in the concentrate, the distribution of the 3,281 metric tons of uranium obtained from conventional mining and ISL in 2004 was as follows (OECD-IAEA, 2005):

- Open-pit and underground mining: 2,880 metric tons
- ISL: 200 metric tons
- Heap leaching: approximately 190 metric tons
- In-place leaching (slope or block): 11 metric tons

During the workshop, it was stated that Russia intends to increase its production of uranium from 4,900 metric tons in 2010 to 18,000 metric tons by 2020 (Shatalov, 2007). Conventional mining of uranium has left large areas (several square kilometers) of land contaminated with mill tailings. For example, Uzbekistan has 2.8 km^2 of such land; Tajikistan, 3 km^2; Kyrgyzstan, 6.5 km^2; Kazakhstan, 2.5 km^2; and Russia, 2 km^2 (Karamushka and Ostroborodov, 2008). As of 1990 the former Soviet Union had generated approximately 5 billion metric tons of mill tailings (Bradley, 1997).

THE WISMUT SITE IN FORMER EAST GERMANY

The major source of uranium for the Soviet Union was the mines, collectively now known as Wismut, in Saxony and Thuringia in the German Democratic Republic (GDR). More than 400,000 people have worked at Wismut, with peak employment reaching 130,000 people in 1950. The total output of the mines was 231,000 metric tons, with peak production of 7,100 metric tons in 1967. During the Cold War, very little was known about uranium mining in the Soviet Union. However, in 1954 a joint Soviet-GDR company, Wismut, was formed. After German reunification, production ceased and Wismut became the property of the Federal Republic of Germany. When this transfer occurred, much previously unavailable data became public. Wismut is responsible for the remediation program to safely isolate and control the enormous quantity of uranium production residue, 312×10^6 m^3 of waste rock, 1,518 ha of waste rock pile area, 161×10^6 m^3 of tailings volume, and a tailings pond area of 724 ha. The estimated cost of cleanup is U.S. $6 billion over 15 years. (More recent information says that the cost will be "in the order of 6.2 billion euros" [Hagen, 2007].) Although working conditions were extremely poor and the environmental contamination was severe, the doses to the critical group were surprisingly small, 0.26 mSv/a, although approximately 5,500 cases of occupational radiation-induced cancer of the lung were identified (Kirchmann and Cigna, 2003).

Some doubt has been expressed about the necessity of such large remediation costs. For example, Jiri Hulka has stated: "In my view, a great deal of remediation work is unnecessary on radiological grounds; it is carried out for political, aesthetic or other reasons. For example, I do not think it was necessary, on radiological grounds, to spend over 10 million euros on the remediation of uranium tailings at Wismut, in Germany" (Hulka, 2003). At a conference in 2004, I also expressed similar reservations about the benefits relative to the cost of the U.S. mill tailings program: "I was recently in Wyoming with Chinese colleagues to visit uranium mining and milling remediation sites. I was struck by the hundreds of millions of dollars being spent to protect a population that might be there in 200 to up to 1,000 years in the future from statistical deaths. I wondered if our intergenerational concerns had not blinded us to our intra-generational concerns . . . . I was further struck by the fact that because of the long term buildup of thorium and radium daughter products that the maximum doses would not occur for hundreds of thousands of years. Does it make sense to spend those amounts of money now to protect these future generations while allowing so many local members of the population to live below the poverty level?" (Parker, 2004).

Some appreciation of the magnitude of the Wismut properties can be gained from Figure 11-1, which shows a small portion of the mining area.

URANIUM RECOVERY IN THE UNITED STATES

Most mines that produced uranium as a primary commodity are, or were, located in Colorado, Utah, Wyoming, New Mexico, and Arizona. They are typically on federal and tribal lands. The number of locations associated with uranium, as identified in the U.S. Environmental Protection Agency (EPA) database, is around 15,000. Of these uranium locations, more than 4,000 are mines having documented production (EPA, 2006). Figure 11-2 shows where most of the uranium mining activities were located in the West.

Currently, 80 percent of mined uranium in the United Stated comes from ISL. Uranium mining in the United States had decreased significantly since the 1970s. However, the price of uranium has soared from $29 per pound in June 2005 to $138 per pound as of June 18, 2007 (*Uranium Miner*, 2007).

While the actual numbers might change, the U.S. Nuclear Regulatory Com-

FIGURE 11-1 A view of a portion of the Wismut facilities.

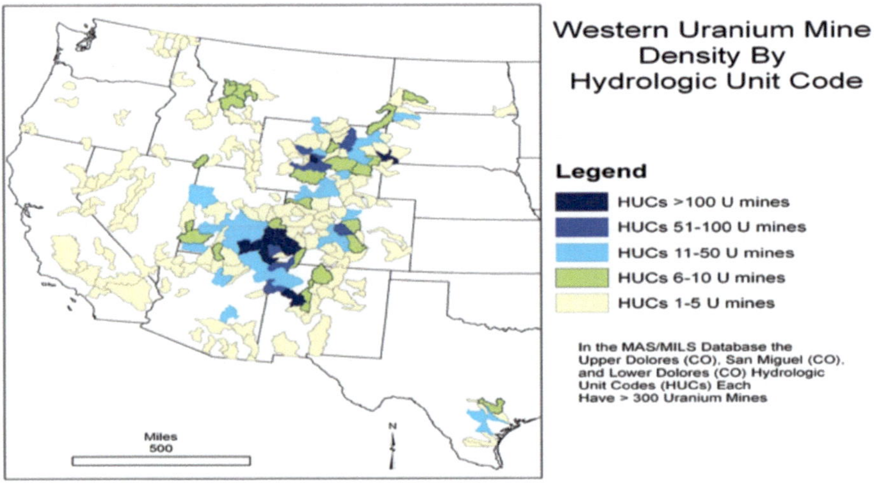

FIGURE 11-2 Location of uranium mines in the western United States.

mission (NRC) is currently expecting 25 licensing actions for uranium recovery in the 2008-2009 time frame. Specifically, these licensing actions include 14 new operations (11 ISL and 3 conventional mines) and 11 restarts, of which 9 are ISL and 2 are conventional mines (USNRC, 2007).

URANIUM MILL TAILINGS REMEDIATION IN THE UNITED STATES: THE URANIUM MILL TAILINGS RADIATION CONTROL ACT (UMTRCA) OF 1978

While UMTs typically contain relativity low activity, they are perceived by many to constitute a serious hazard based on their very large volume. The Uranium Mill Tailings Radiation Control Act (UMTRCA) of 1978 was passed in response to a perceived need for the implementation of engineered controls to prevent releases from UMT sites. Title I of the act addresses active and inactive facilities at the time the act was passed that had recovered uranium for the federal government. The U.S. Department of Energy (DOE) was given the responsibility of remediating these sites under a general license from the NRC, and the EPA was charged with developing protective standards for UMT sites.

Title II of UMTRCA addressed UMT remediation at commercial facilities active in 1978 and future facilities licensed by the NRC. As of 2007, there are 16 uranium recovery facilities licensed by the NRC—12 conventional mines and 4 ISL. Closure and remediation of Title II sites are the responsibility of the licensee (USNRC, 2006).

Table 11-1 provides information about Title II sites currently undergoing decommissioning.

TABLE 11-1 Title II Sites Undergoing Decommissioning

Name	Location	Estimated Decommissioning Costs ($)
American Nuclear Corporation	Casper, WY	3.2 million
Bear Creek	Converse County, WY	900,000
COGEMA Mining Inc.	Mills, WY	12.1 million
ExxonMobil Highlands	Converse County, WY	
Homestake	Grants, NM	55.5 million
Pathfinder-Lucky MC	Gas Hills WY	
Pathfinder-Shirley Basin	Shirley Basin, WY	
Rio Algom-Ambrosia Lake	Grants, NM	18 million
Sequoyah Fuels Corporation	Gore, OK	
Umetco Minerals Corp.	East Gas Hills, WY	
United Nuclear Corporation	Church Rock, NM	3.7 million
Western Nuclear Inc.-Split Rock	Jeffrey City, WY	

SOURCE: USNRC, no date.

Information about the status of Title I sites is maintained by DOE through the Office of Legacy Management in Grand Junction, Colorado (DOE, 2001). The NRC Uranium Recovery Licensing Branch, Decommissioning and Uranium Recovery Licensing Directorate, Division of Waste Management and Environmental Protection, Office of Federal and State Materials and Environmental Programs, maintains records on Title II sites.

UMTRCA designated the federal government (DOE) as the long-term custodian for all sites remediated under Title I. For Title II sites, the host state could assume the role of long-term custodian. As of 2007, no host states have come forward, and DOE has taken the responsibility for custodial care of all sites regulated under UMTRCA.

At this time, all but 2 of the approximately 20 Title I sites have been remediated. At one location (the Moab, Utah, site), the tailings are close to the banks of the Colorado River, and DOE has agreed to move them to a safer location at a cost of several hundred million U.S. dollars. The remaining site at Grand Junction, Colorado, will remain open to receive additional tailings for several more years. All but two of these sites are in semiarid climates. Canonsburg and Burrell, Pennsylvania, are in humid environments.

Surveillance and monitoring of closed UMT sites are conducted through the DOE Office of Legacy Management in Grand Junction, Colorado, for both Title I and Title II sites. Inspection reports are available through the DOE Office of Legacy Management (DOE, 2005b, no date [b]).

Through 1999 the total cost of activities for the Title I sites was approximately $1.5 billion (DOE, no date [a]). Costs for the remediation of Title I sites had been estimated by DOE, prior to the passage of UMTRCA in 1978, to be in the range of $150 million-200 million (GAO, 1995).

RISKS ASSOCIATED WITH UMT SITES

As indicated above, the sheer volume of waste at UMTs, despite the relatively low activity associated with the tailings wastes, has engendered a perception that UMT sites pose a serious hazard and potential risk to human health and the environment. The actual risk, of course, depends upon the nature and amount of radioactive materials to which the public can be exposed.

The U.S. National Research Council examined the scientific basis for risk assessment at UMT sites in 1986. Although there are many National Research Council reports on topics that deal with aspects of management of uranium mill tailings, this is the only report that approaches the problem holistically (National Research Council, 1986). In its report the authoring committee states the following:

- Surveillance should be maintained to take necessary corrective action if needed.

- Risk management strategies must be site specific.
- The health risks posed by radon differ widely depending upon distance of potential receptors from the site.

ENGINEERED APPROACHES TO THE CLOSURE OF UMT SITES

Closures of inactive UMT sites in the United States have employed the use of engineered barriers (surface covers) whose primary functions are the prevention of the release of radon gas to the atmosphere, the infiltration of precipitation and subsequent leaching of the waste and transport of waste constituents to groundwater, and the transport of waste materials to surface water through runoff.

These covers typically include a primary radon release and infiltration prevention barrier of compacted clay soils, with other layers included to protect the primary barrier. Figure 11-3 shows the cover design employed at Durango, Colorado.

The function of the cover is described for the 360-acre Cheney site, which contains the 98-acre disposal cell and is located about 18 miles southeast of Grand Junction. The cell is about 80 feet deep from lowest to highest points and is capped with an engineered, 7-foot-thick multicomponent cover. The cover is composed of a 1.5-foot-thick transition layer placed directly on the radioactive materials; a 2-foot-thick clay layer serving as the radon barrier and minimizing water infiltration; a 2-foot-thick layer of clay above the radon-infiltration barrier minimizing freeze-thaw damage; a 6-inch-thick coarse-grained bedding layer covering the freeze-thaw barrier to minimize capillary movement of fluids and

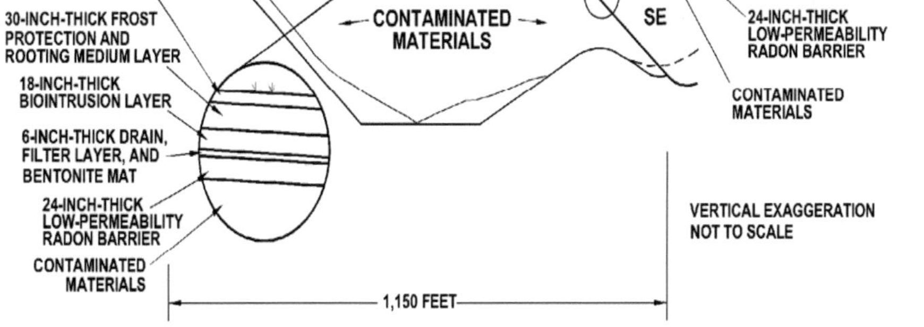

FIGURE 11-3 Mill tailings cover design, Durango, Colorado.

provide surface drainage; and a 12-inch-thick riprap erosion protection layer (DOE, 2005a). Several Title I sites also had existing groundwater contamination, and groundwater management measures were necessary as well.

PERFORMANCE ISSUES AND MONITORING

When considering the design approach for an engineered cover at a UMT site, local climate and environmental factors are important, as the cover design is dependent upon rainfall and infiltration rates. At some sites, however, the engineered covers are experiencing degradation as a result of natural processes, especially erosion and root intrusion (biointrusion). For example, at the Title I site in Burrell, Pennsylvania, Japanese knotwood trees penetrated the primary barrier just a few years after the site was closed and increased the hydraulic conductivity of the barrier by two orders of magnitude over the design value. A risk assessment, however, revealed that the biointrusion process had increased evapotranspiration and that corrective action was not required (Waugh, 1999). Examples of erosion have also been seen on rock-covered side slopes, for instance, at the Durango, Colorado, site. Maintenance has not yet been deemed necessary but may be required in the future (DOE, 2006).

In its report on the potential hazards associated with UMT sites and their closure, the International Atomic Energy Agency (IAEA, 1997) stated the following:

> In the long term, typical scenarios or occurrences needing attention are intrusion and erosion (i.e., human or natural-caused degradation of engineered barriers, etc.).
>
> - Maintaining institutional control for long time periods.
> - Upgrading long-term confinement systems (e.g., upgraded near-surface barriers, disposal into a lake, or by backfill into an underground mine).
> - Implementing advanced solutions where practicable and economically feasible, such as extracting selected radioisotopes from the tailings.
> - Selecting dispersion instead of containment if a safety assessment case allows it.
> - Comparing "enhanced activity" conditions with site or local "natural activity."
> - Adopting less conservative or probabilistic scenarios or assumed occurrences.
> - Effects of radioactive and non-radioactive substances on the environment.

Recently, another National Research Council committee examined issues concerning the long-term performance of engineered barriers (National Research Council, 2007). The following comments and recommendations are taken from its report:

- "Given that development of optimal designs for lifetimes of thousands of years is likely to be both infeasible and prohibitively expensive, designs that allow for recovery, repair, and/or replacement are to be encouraged" (p. S2).
- "Tomographic imaging and seismic velocity surveys . . . . Multispectral imaging . . . . Interferomagnetic synthetic aperture radar, light detection, and ranging and other airborne/satellite techniques . . . . However, to date, these technologies have yielded little data that can be used to quantitatively and reliably monitor barrier systems" (p. S3).
- "The estimated service lives of geomembranes decrease from 1,000 years at 10°C to only about 15 years at 60°C" (p. S3).
- "Cover systems are effective at isolating waste, as long as periodic maintenance is performed" (p. S4).
- "Recommendation 5: Regulatory agencies and research sponsors should support the validation, calibration, and improvement of models to predict the behavior of containment system components and the composite system over long periods of time. These models should be validated and calibrated using the results of field observations and measurements."

Closed UMT sites in the United States are monitored annually, including physical inspections of cover integrity and access restrictions such as signs and fences. While a requirement for perpetual monitoring and maintenance of these sites is unrealistic, it is noted that the general NRC license contains no expiration date. Research is ongoing to develop improved designs that can better accommodate natural processes and environmental change.

A summary of the conventional sites being remediated and their costs is presented in Table 11-2, and the costs for decommissioning nonconventional sites, ISL, are given in Table 11-3.

TABLE 11-2 Remediation of UMTRCA Title I Uranium Mill Sites Under the UMTRA Project Summary Table: Uranium Ore Processed, Disposal Cell Material, and Cost for Remediation as of December 31, 1999

	Uranium Ore Processed		Disposal Cell	Remediation Project Cost			
Remediation Project (Mill Site Name, State)	Ore (Million Short Tons)	Uranium Production (Million Pounds U_3O_8)	Remediated Material Volume (Million Cubic Yards)	Total Cost (Thousand U.S. Dollars)	Per Pound Produced (Dollars per Pound U_3O_8)	Per Unit of Remediated Material (Dollars per Cubic Yard)	Per Unit of Radiation Avoided (Dollars per Curie, Ra-226)
Minimum Cost per Curie Avoided Edgemont, SD	1.98	6.86	3.00	5,411	0.79	1.80	10,267.55
Maximum Cost per Curie Avoided Lowman, ID	0.20	0.37	0.13	18,434	50.47	142.46	1,536,166.67
Total[a] and Averages	27.17	116.53	46.07	1,476,340	12.67	32.04	105,249.88

[a]For 24 sites.
SOURCE: DOE, 2005a.

TABLE 11-3 Estimated Decommissioning Costs for U.S. Nonconventional Uranium Production Facilities (in 1994 dollars)

Name and Costs	Well Field Restoration Costs ($, thousands)	Groundwater Restoration Costs ($, thousands)	Other Costs ($, thousands)	Total Costs ($, thousands)	Groundwater Restoration Costs (% of Total)
Maximum: Burns Ranch/ Clay West	3,808	15,994	15,211	35,013	46
Minimum: Tex-1	201	176	199	576	31
Average	931	2,819	3,233	7,032	
Totals[a]	13, 027	39, 460	45, 961	98, 448	40

[a]For 14 sites.

SOURCE: DOE, 1995, p. 38; Davis and Curtis, 2007, p. 15.

REFERENCES

Bradley, Don J. 1997. Behind the Nuclear Curtain: Radioactive Waste Management in the Former Soviet Union. Columbus, Ohio: Battelle Press.

Davis, J. A., and G. P. Curtis. 2007. Consideration of Geochemical Issues in Groundwater Restoration at Uranium In-Situ Leach Mining Facilities, NUREG/CR-6870, U.S. Nuclear Regulatory Commission, January.

DOE (Department of Energy). No date (a). Energy Information Administration. Uranium Mill Sites under the Uranium Mill Tailings Remedial Action (UMTRA) Project. Available online at *www.eia.doe.gov/cneaf/nuclear/page/umtra/title1sum.html.*

DOE. No date (b). Office of Legacy Management in Grand Junction, Colorado. Available online at *www.lm.doe.gov/land/sites/co/gj/gjo/gjo.htm.*

DOE. 1995. Energy Information Administration. Decommissioning of U.S. Uranium Production Facilities. DOE/EIA-0592. Available online at *tonto.eia.doe.gov/ftproot/nuclear/0592.pdf.*

DOE. 2001. Grand Junction Office. Guidance for Implementing the Long-Term Surveillance Program for UMTRCA Title I and Title II Disposal Sites. GJO-2001-TAR, April.

DOE. 2005a. Energy Information Administration. Grand Junction (Climax Uranium) Mill Site. Available online at *www.eia.doe.gov/cneaf/nuclear/page/umtra/grandjunction_title1.html.*

DOE. 2005b. Office of Legacy Management. Annual Site Inspection and Monitoring for Uranium Mill Tailings. Radiation Control Act Title I Disposal Sites, DOE Office of Legacy Management DOE-LM/GJ1015—2005, December.

DOE. 2006. Office of Legacy Management. Durango, Colorado, Processing and Disposal Sites. Available online at *www.lm.doe.gov/documents/sites/co/dur_d/fact_sheet/Durango.pdf.*

EPA (Environmental Protection Agency). 2006. Office of Radiation and Indoor Air Radiation Protection Division. Uranium Location Database Compilation, EPA 402-R-05-009. Available online at *www.epa.gov/radiation/docs/tenorm/402-r-05-009.pdf.*

GAO (General Accounting Office). 1995. Uranium Mill Tailings: Cleanup Continues, but Future Costs are Unknown, GAO/RCED-96-37. Washington, D.C.: GAO. Available online at *www.gao.gov/archive/1996/rc96037.pdf.*

Hagen, M. 2007. The Wismut uranium mining rehabilitation project running for 15 years—lessons learned and achievements. Advanced Materials Research (20-21):243-247.

Hulka, J. 2003. Issues and Trends in Radioactive Waste Management. Proceedings of an International Conference, Vienna, December 9-13, 2002. Vienna: International Atomic Energy Agency, pp. 405-406. Available online at *www-pub.iaea.org/MTCD/publications/PDF/Pub1175_web/Book/Pub1175_web.pdf*.

IAEA (International Atomic Energy Agency). 1997. Waste Technology Section. Closeout of Uranium Mines and Mills: A Review of Current Practices, IAEA-TECDOC-939. Vienna: IAEA.

Karamushka, V. P., and V. V. Ostroborodov. 2008. Lands Damaged as a Result of Uranium Ore Mining Operations in the Russian Federation. Pp. 61-68 in Cleaning Up Sites Contaminated with Radioactive Materials: International Workshop Proceedings. Washington, D.C.: The National Academies Press. Presented at the International Workshop on Cleaning Up Sites Contaminated with Radioactive Materials, Moscow, Russian Federation, June 4-6, 2007.

Kirchmann, R. J. C., and A. A. Cigna. 2003. Radioactivity from Military Installations Sites and Effects on Population Health, SCOPE-RADSITE. Authors of the Wismut section were K. Wichterey and P. Schmidt.

National Research Council. 1986. Scientific Basis for Risk Assessment and Management of Uranium Mill Tailings. Washington, D.C.: National Academy Press.

National Research Council. 2007. Assessment of the Performance of Engineered Waste Containment Barriers. Washington, D.C.: The National Academies Press.

OECD-IAEA. 2005. Uranium 2005: Resources, Production and Demand. Paris: Organization for Economic Cooperation and Development.

Parker, F. L. 2004. "Rethinking" low-level radioactive waste disposal. Proceedings of the WM'04 Conference, Abstract no. 46151, February 29-March 4, 2004.

Shatalov, V. V. 2007. Environmental Protection Aspects of the Federal Target Program: The Uranium of Russia. Presented at the International Workshop on Cleaning Up Sites Contaminated with Radioactive Materials, Moscow, Russian Federation, June 4-6, 2007.

Uranium Miner. 2007. Available online at *www.uraniumminer.net*. Accessed June 18, 2007.

USNRC (U.S. Nuclear Regulatory Commission). No date. Locations of Uranium Recovery Sites Undergoing Decommissioning. Available online at *www.nrc.gov/info-finder/decommissioning/uranium*.

USNRC. 2006. Fact Sheet on Uranium Mill Tailings. Available online at *www.nrc.gov/reading-rm/doc-collections/fact-sheets/mill-tailings.html*.

USNRC. 2007. National Mining Association. Uranium Recovery Workshop, Denver, Colorado, May 14-16, 2007.

Waugh, W. J. 1999. Plant Encroachment on the Burrell, Pennsylvania, Disposal Cell: Evaluation of Long-Term Performance and Risk, DOE Grand Junction Office, GJO-99-96-TAR.

World Nuclear Association. 2007. Available online at *www.world-nuclear.org/education/mining.htm*.

12

Experience in Rehabilitating Contaminated Land and Bodies of Water Around the Mayak Production Association*

Yu. V. Glagolenko, Ye. G. Drozhko, and S. I. Rovny,
Mayak Production Association

THE CREATION OF ENVIRONMENTAL PROBLEMS AT THE MAYAK PRODUCTION ASSOCIATION

Since the moment of its creation in the late 1940s, the enterprise known today as the Mayak Production Association—Federal State Unitary Enterprise has faced scientific-technical and industrial tasks of unprecedented complexity in connection with its work on building nuclear weapons. Over the course of decades, the achievement of military-political goals pushed environmental protection questions into the background, which ultimately led to serious radioecological problems.

The current radioecological situation in the region where Mayak is located arose as a result of the following factors:

- Discharges of liquid radioactive wastes into the open hydrographic system of the Techa River (1949-1956)
- The 1957 accident at a liquid radioactive waste holding tank, which resulted in the creation of the East Urals Radioactive Trace

*Translated from the Russian by Kelly Robbins.

- The use of the V-9 (Karachai) and V-17 (Staroye Boloto) reservoirs for the storage of medium-level liquid radioactive wastes
- The creation of the man-made reservoirs of the Techa Cascade for the storage of low-level liquid radioactive wastes
- Windborne dispersal of radioactive sediments from the exposed shoreline of the Karachai Reservoir (1967)

The fundamental problems associated with Mayak's ongoing activities are linked to the use for technical purposes of eight industrial reservoirs at the enterprise to store liquid radioactive wastes accumulated as a result of defense program operations (see Figure 12-1).

To ensure that thorough and comprehensive measures are taken to make Mayak's production activities environmentally safe, a comprehensive plan has been developed to address environmental problems associated with the enterprise's current and past operations.

The essence of this effort lies in implementing the following four-part program:

1. Reducing and ultimately halting discharges of all liquid radioactive wastes into industrial reservoirs
2. Eliminating the most radiologically hazardous reservoirs—Karachai (V-9) and Staroye Boloto (V-17)
3. Ensuring the safe operation of the Techa Cascade of reservoirs
4. Reducing the volume and radioactivity level of the high-level wastes stored in holding tanks

Each of these problems merits separate consideration. The problem of the Techa Cascade is of the highest social significance.

ENSURING THE SAFE OPERATION OF THE TECHA CASCADE

Problems associated with reservoir safety, particularly for the reservoirs of the Techa Cascade, have been substantially exacerbated by changes in climate conditions in the region. Since the early 1980s, the region has experienced wetter conditions (disparity between annual precipitation and evaporation from bodies of water), and as a result, water levels in most of the reservoirs, including those of the Techa Cascade, have approached regulatory maximums.

For example, an analysis of water balance components for Reservoir V-11, the final reservoir in the cascade, indicates that the primary reason for the rise in its level since 1980 is the change in meteorological conditions in the region (see Table 12-1). Whereas from 1950 until the 1980s, the average level of evaporation in the Mayak region exceeded precipitation by 100 mm per year, the situation

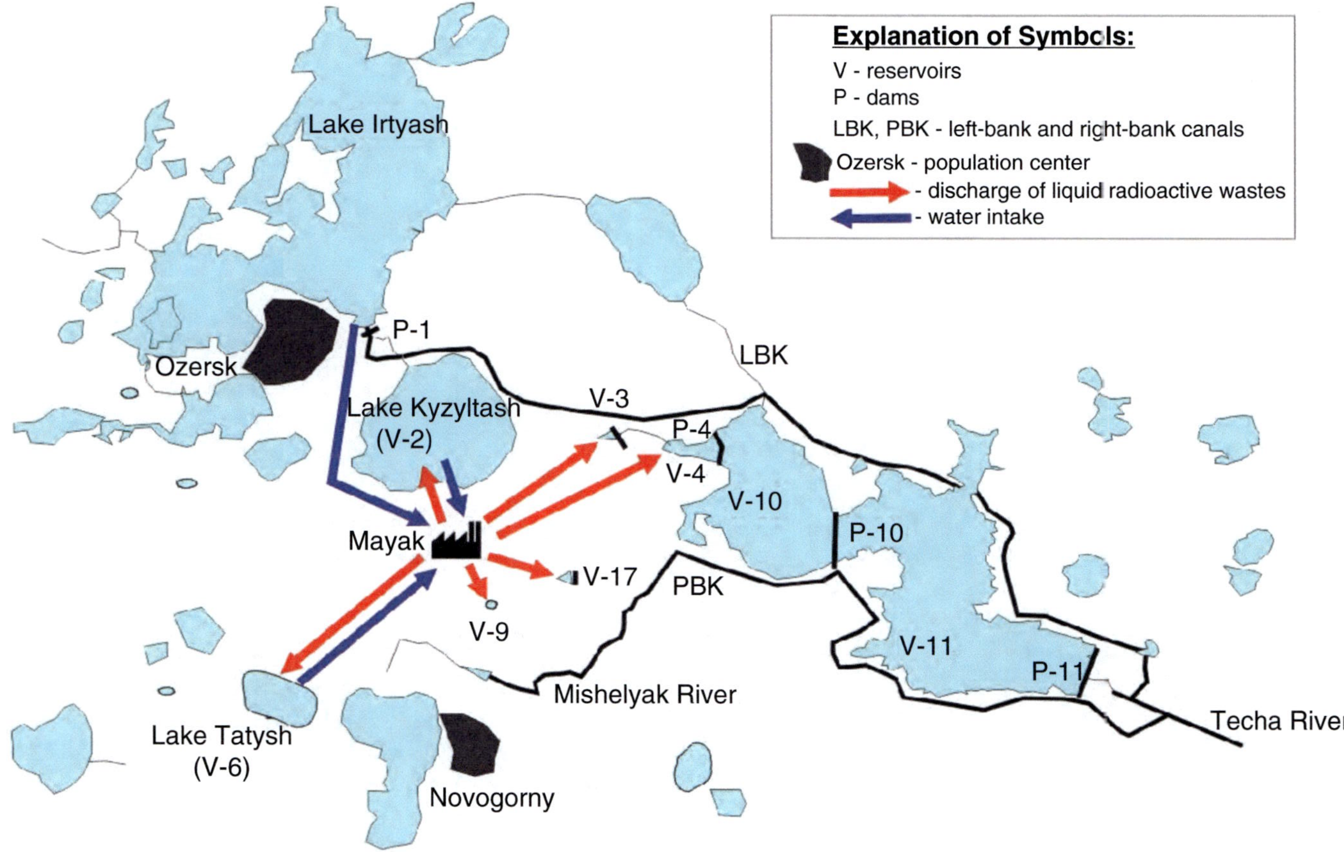

FIGURE 12-1 Water supply and discharge system at Mayak.

TABLE 12-1 Water Balance in the Techa Cascade Reservoirs, 1993-2006

	Change in Volume	
Water Balance Component	Million m^3	%
Total volume, Techa Cascade reservoirs		
as of January 1, 1993	300.64	—
as of January 1, 2007	330.81	—
Change in volume	30.17	5.4
Inputs:		
atmospheric precipitation	439.2	79.1
household and industrial wastewater discharges, storm drain runoff	83.4	15.0
discharges of low-level wastes	4.5	0.8
underground and surface flows from drainage basin	28.0	5.0
Total inputs	555.1	100
Outputs:		
surface evaporation	327.8	59.1
seepage from Techa Cascade reservoirs	197.1	35.5
Total outputs	524.9	94.6

was reversed from 1980 to 2006, when precipitation exceeded evaporation by an average of 90 mm per year.

The situation in the Techa Cascade became extremely critical in 1999-2000, when the water level in Reservoir V-11 rose 1.2 m in 1.5 years. The recurrence of such a situation in the future cannot be ruled out.

In the next 50 years, plans call for two main periods of rehabilitation of the Techa Cascade in the course of its operations:

- During the first period, which will last 6 to 8 years, volumes of liquid waste discharges into the reservoirs will be reduced, and the water balance in the Techa Cascade will be stabilized.
- During the second period, conditions are to be created for the long-term, controlled, and safe storage of the liquid wastes that have accumulated in the Techa Cascade reservoirs, and technical solutions and projects will be implemented to reduce water levels to acceptable standards.

The rise of the water level in Reservoir V-11 will lead to increased infiltration of strontium-90 into the canals and ultimately into the Techa River (see Figure 12-2).

Several measures were taken in 1999-2006 to stabilize the water level in the reservoirs of the Techa Cascade and reduce the influx of radioactive substances from the cascade into the Techa River system:

- The northern borehole water extraction system was put into operation, making it possible to remove up to 1.5 million m^3 of groundwater from the Techa Cascade.

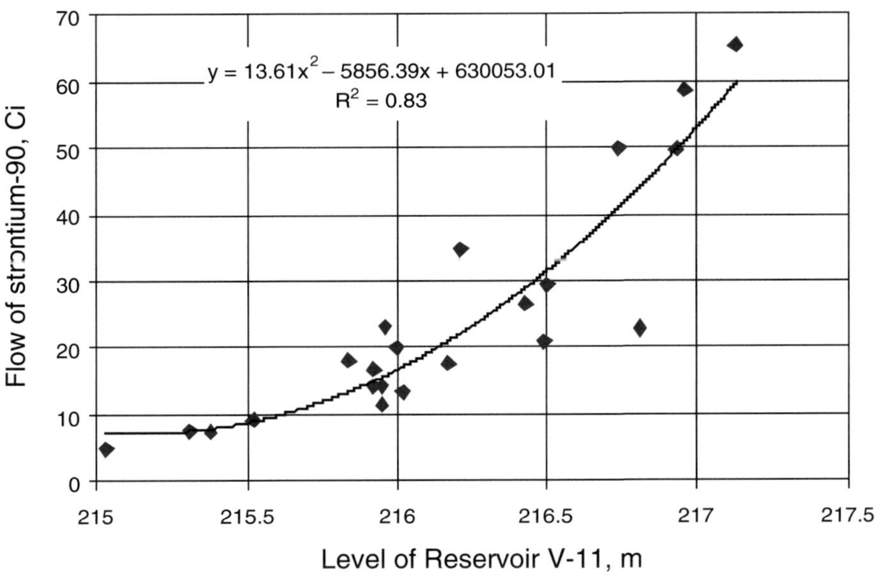

FIGURE 12-2 Correlation between total infiltration of strontium-90 into the surface levels of the Techa River and water level in Reservoir V-11.

- The water control capabilities of the Right Bank Canal were restored by ridding it of ash washed into the canal from the Argayash power plant. The volume of ash extracted from the canal totals 15,000-17,000 m^3 annually. The existing system of upland canals (Left Bank and Right Bank) in the Techa Cascade facilitates the transfer of surface and groundwater in the cascade's catchment basin.
- Plans have been developed and work has been initiated to create a primary and comprehensive sewerage system to transfer decontaminated water into the open hydrographic network. Creation of such a system is the most significant measure in stabilizing the water level in Reservoir V-11. Completion of the project will make it possible to reduce the level of water flowing into the Techa Cascade by 3 million m^3 per year.
- Measures have been taken to reduce discharges of liquid radioactive wastes.

Efforts to reduce the input side of the water balance may be of little effect if increasingly wet conditions continue. Therefore, a great deal of attention is being devoted to improving the stability of Dam P-11, which is a key element in the system of hydrotechnical structures in the Techa Cascade that ensure the safe operation of the entire reservoir system.

Built in 1964 based on a design by the Kuibyshev Branch of the Hydroproject Institute, the dam at Reservoir V-11 (P-11) is of a low-pressure earthen-fill type. Dam P-11 was rehabilitated in 1975 and 1991, which made it possible to operate it right up to the 217.32 m mark.

Exploration and testing of the physical-mechanical properties of the soils making up the body of the dam indicated the presence of weakened zones in the upper part of the structure. It was decided to install an additional antiseepage element deep within the body of the dam. The most effective measure for preventing infiltration and developing suffusion processes in the body of the dam is the creation of an antiseepage curtain, a "wall in the soil." An antiseepage lock will be created in the central part of the dam by driving a series of metallic plates (Larsen plates) into the ground.

Work to sink the foreshaft and experimental work to construct the wall-in-the-soil structural element itself were successfully completed in 2006. This year (2007), plans call for completing work to create the antiseepage curtain in the body of Dam P-11. Constructing such antiseepage elements will make it possible to operate Dam P-11 more safely and bring it up to first-class level with regard to static, seismic, and infiltration stability parameters. Most importantly, this will improve its operating capacity, that is, its normal level of resistance.

Fundamental resolution of the Techa Cascade problem may be achieved only by selecting and implementing a plan for artificially lowering the water level in Reservoir V-11. One option that has not been implemented would be to create a complex for water purification and transfer of processed water from Reservoir V-11 into the Left Bank Canal. The enterprise has adopted and tested a membrane-sorption water purification system for Reservoir V-11 that would make it possible to discharge the purified water into the open hydrographic system. This system has processed more than 300 m^3 of real water (with strontium-90 activity of 1,500 Bq/L). The purification coefficient achieved was more than 300, which meets requirements and radiation safety standards adopted in the Russian Federation.

THE KARACHAI RESERVOIR (V-9)

Taking Reservoir V-9 out of operation requires the complete cessation of its use as a receiving basin for liquid radioactive wastes. Basic data on this reservoir are presented in Table 12-2.

After the decision was made to eliminate Reservoir V-9 and experimental work was carried out, planned efforts have proceeded since 1986 to drain and fill in this body of water as part of projects for eliminating and decommissioning the reservoir in the first, second, and third stages. Work to eliminate (decommission) the Karachai Reservoir is being carried out by filling in the basin with rock. In areas where highly radioactive technogenic sludge is found, special solid concrete blocks in reinforced sections are also put in place using specialized radiation-shielded equipment.

TABLE 12-2 Basic Parameters of Reservoir V-9

Parameter	Value
Area, km^2	0.11
Volume, million m^3	0.4
Activity level, million Ci[a]	120
Distribution of activity, %	
Water	5
Sediments	95

[a]Activity due to strontium-90 and cesium-137.

The first stage of work entailed filling in the northeastern part of the reservoir and building a series of dams to divide the body of water into sections. Filling in the northeastern section made it possible to localize most of the radioactive technogenic sludge that had formed in the reservoir, which contained the bulk of the accumulated radioactivity. Completing this stage was the most complex part of the entire project, inasmuch as it involved working in conditions of extremely high levels of radiation with sediments constantly coming to the surface as dam fill was being laid. The successful filling of the northeastern section of the reservoir fundamentally improved the radiation situation in the area surrounding Reservoir V-9 and the radiochemical plant. This stage was successfully completed in 1990. As a result of this work, about 60 percent of mobile sediments by volume were locally contained, along with 70 percent of all radionuclides accumulated in the reservoir. Completion of work in the first phase of the project to eliminate Reservoir V-9 proved that all technical solutions adopted were well founded, demonstrated that the task in general could realistically be completed, and made it possible to move forward on efforts to fill in the entire reservoir.

At present, hydrological conditions in the Karachai Reservoir basin have changed as a result of changes in meteorological conditions and the increased water levels of recent years. This situation has required the correction of design solutions and the development of a third phase in the project to decommission Reservoir V-9. These plans were completed by the All-Russian Design and Scientific-Research Institute of Comprehensive Energy Technology in 2004. The plans call for filling in the remaining part of the reservoir while simultaneously gradually stopping discharges of liquid radioactive wastes. This work should result in elimination of the reservoir and recultivation of the land, but long-term monitoring, control, and maintenance will be needed.

From the standpoint of the Karachai Reservoir's impact on the environment, groundwater contamination remains a rather pressing problem. Forecasts of how the situation may develop over a fairly long period (300 years) indicate that in the future there will be practically no radiologically significant discharge of contaminated groundwater into the open hydrographic network.

STAROYE BOLOTO RESERVOIR (V-17)

Eliminating the Staroye Boloto Reservoir (V-17, described in Table 12-3) will involve use of the fill-in technology tested and used in the closure of Reservoir V-9. Furthermore, experimental work on covering over sediments was carried out at Reservoir V-17 in 2004-2005, which made it possible to confirm basic technical solutions involved in reservoir decommissioning. After the completion of work to fill in the reservoir, the existing radioactive waste storage site will be operated as a near-surface solid radioactive waste burial site.

TABLE 12-3 Basic Parameters of Reservoir V-17

Parameter	Value
Area, km^2	0.13
Volume, million m^3	0.36
Activity level, million Ci[a]	1.2
Distribution of activity, %	
Water	0.5
Sediments	99.5

[a]Activity due to strontium-90 and cesium-137.

REDUCING DISCHARGES INTO INDUSTRIAL RESERVOIRS THROUGH ORGANIZATIONAL-TECHNICAL MEASURES AND WATER USE OPTIMIZATION

With the aim of gradually reducing discharges, organizational-technical measures were carried out before the unit for cleaning and solidifying liquid radioactive wastes began operation. These measures made it possible to reduce discharges of medium-level radioactive wastes into Reservoirs V-9 and V-17 by 2,260 m^3, of low-level technological wastes into the Techa Cascade by 73,000 m^3, and of low-level nontechnological wastes by 410,000 m^3. Further discharge reductions are possible only by optimizing technological processes at the enterprise. To achieve this goal, plans call for modernizing certain process stages and equipment and partially replacing the outdated technology.

As a result of a number of scientific-practical and experimental design efforts, the two most promising technological plans were selected for optimizing spent fuel reprocessing. The plans chosen do not contradict the existing technology, but take into account its geography and the existing equipment and technological linkages among the various production divisions at the plant. They also make it possible to reduce the volume of medium-level radioactive waste created during spent fuel reprocessing by about half.

LIQUID RADIOACTIVE WASTE REPROCESSING

Completely eliminating discharges into special reservoir-repositories as part of plant modernization requires creating a unit to reprocess liquid wastes.

Management of High-Level Radioactive Wastes

Since 1987, Mayak has reprocessed high-level radioactive wastes in direct electric heating furnaces using a technology that turns the wastes into sodium aluminum phosphate glass using an EP-500–type furnace. While the complex has been in operation, it has reprocessed 20,000 m^3 of high-level wastes with a total activity of more than 460 million Ci of β-emitting radionuclides, and more than 4,000 metric tons of glass has been produced. The storage facility is 59 percent full. The complex is processing a wide range of wastes from current production operations as well as some that accumulated previously. Waste processing volumes are presented in Table 12-4.

TABLE 12-4 Quantity of Wastes Reprocessed

Volume of Liquid Wastes Processed, m^3	Quantity of Radionuclides Processed, kCi		Quantity of Glass Produced, metric tons
	α Emitters	β Emitters	
20,000	5,830	458,400	4,139

Before the creation of new-generation remote furnaces, the primary method of reprocessing and solidifying high-level wastes remains their vitrification in direct heating EP-500–type furnaces. In 2006, construction was completed on the EP-500/4 furnace and the unit was put into operation. The expected service life of the furnace will end in 2009; therefore, to ensure that the technological process for high-level waste reprocessing is uninterrupted and to provide for the storage of the vitrified wastes, appropriate plans have been made to create the next in the series of electric furnaces.

Management of Medium-Level Wastes

At present, medium-level radioactive wastes that are created are discharged into industrial reservoirs V-9 and V-17; therefore, the problem of taking the reservoirs out of operation and decommissioning them is directly linked with the problem of halting the discharge of medium-level wastes into them.

In 2006 the All-Russian Design and Scientific-Research Institute of Comprehensive Energy Technology developed the design for a medium-level radioactive waste cementing complex at the radiochemical plant ("Creation of a Complex for

Cementing Liquid and Heterogeneous Medium-Level Wastes"). The design plans identify a technological setup for medium-level waste reprocessing, including a standardization of the entire range of medium-level wastes, single-stage steaming of solutions, and cementing and burial in a near-surface–type repository using sectional-poured concrete technology.

Management of Accumulated High-Level Wastes

During the years of operation of Mayak's radiochemical plants, more than 29,000 m^3 of high-level liquid wastes with a total activity of 366 million Ci were accumulated in repository vessels. These wastes included the following:

- Aluminum, chromium, and iron hydroxide pulps and nickel ferrocyanide created after the purification of alkaline decantates in the acetate precipitation technology for the separation of weapons-grade plutonium
- Complex salt nitrate solutions, primarily refined products formed as a result of extraction processing of weapons-grade plutonium

The storage of highly active suspensions and solutions in steel storage vessels is seen as one stage in waste reprocessing that reduces their activity as a result of the radioactive decay of short-lived radionuclides. The wastes can be subsequently reprocessed at reduced cost and effort. The duration of storage is determined by the expected service life of the storage vessels. For the first vessels, which were put into use in 1968, this period ends in 2018, which means that these wastes need to be moved into safer conditions.

Currently existing high-level waste management practices at the enterprise's radiochemical plant RT-1 involve vitrifying a mixture of radionuclides and accompanying stable chemical additives and then holding the glass blocks for temporary controlled storage in a special repository. However, such an approach is ineffective for accumulated high-level wastes, inasmuch as it is substantially complicated by well-known shortcomings in the direct vitrification method for the following reasons:

- Large volumes of ballast material (salts) are subject to vitrification along with radionuclides.
- The number of macrocomponents in the accumulated high-level wastes has a negative impact on the parameters of the vitrification process.

These shortcomings may be eliminated by implementing a promising method for high-level waste reprocessing that calls for the preliminary separation of certain fractions of active components from the ballast mass of the radioactive wastes, which makes it possible to solidify the latter using cheaper methods such as cementing.

In summary, following are the enterprise's top priority problems in resolving environmental problems:

- Ensuring safe operation of the special industrial reservoirs
- Reducing water balance inputs at the special industrial reservoirs
- Improving the stability of Dam P-11
- Modernizing the radiochemical plant to reduce the amount of liquid radioactive wastes that it creates
- Creating a new phase of the high-level waste vitrification complex
- Creating a complex for cementing medium-level wastes so that discharges of medium-level liquid wastes into reservoirs V-9 and V-17 may be completely halted
- Carrying out work to remove reservoirs V-9 and V-17 from operation and rehabilitate their sites
- Developing technology to put previously accumulated high-level wastes into a form safe for long-term storage

Accomplishing these objectives will facilitate a significant reduction in the environmental impact of Mayak's ongoing production operations and will minimize the effect of factors created by the enterprise's previous activities.

13

Rehabilitation of Contaminated Groundwater Layers Near the Mayak Enterprise Using Deep Burial Technology*

V. G. Skidanov, Ye. N. Kamnev, and A. I. Rybalchenko, Federal State Unitary Enterprise—All-Russian Research, Design, and Surveying Institute of Industrial Technology

Contamination of near-surface underground water (groundwater) occurs at many nuclear industry enterprises in areas where there are surface radioactive waste repositories and at production complex sites. Groundwater contamination has also been noted at the Mayak enterprise, which could create a real threat to surface bodies of water and to surface and groundwater supply sources.

A radical means of cleaning up contaminated groundwater is to set up a system whereby water is pumped out and then cleaned using sorption, membrane, and similar technologies. This produces clean or conditionally clean water, which is subsequently discharged or reused, as well as regenerators or concentrated solutions containing radioactive substances and salts, which are processed into solid form. The use of such a technology requires significant material expenditures, as the volumes of water that must be pumped to produce a significant effect will be in the several hundreds or even thousands of cubic meters per hour.

At the nuclear industry enterprise Chepetsk Mechanical Plant (city of Glazov, Republic of Udmurtia), research was conducted on areas of salt-contaminated groundwater from an alluvial layer near a tailings repository. This research laid the scientific foundations for a geoinfiltration and geomigration model of the up-

*Translated from the Russian by Kelly Robbins.

permost underground layers, the study of which made it possible to propose a system for rehabilitating a layer by pumping out contaminated water. Such measures are necessary because the alluvial layer drains into and possibly contaminated the Cheptsa River. The magnitude of seepage losses from the tailings repository was calculated, broken down by elements of the water balance and including three calculation stages: (1) calculation of initial data, (2) calculation of balance inputs, and (3) calculation of balance outputs. The maximum level of seepage losses from the tailings repository totaled approximately 300,000 m^3 per year.

The overall approach to rehabilitating the water-bearing alluvial layer is to prevent this layer from soaking up contaminated solutions caused by seepage from the tailings repository and subsequently to pump the contaminated water out of the area of its distribution and return it to the tailings repository. Various options for capturing the contaminated water were considered: installation of an antiseepage "wall-in-the-ground" barrier around the perimeter of the tailings repository and the pumping off of contaminated water; installation of an impenetrable barrier around the tailings repository (using the Sergeev method, Moscow State University); and reconstruction of the existing horizontal drainage system. All of these measures are aimed at limiting the influx of water from the tailings repository into the alluvial layer, but they do not address matters of how to clean the contaminated layer drained by the Cheptsa River or how to handle the contaminated water that is pumped out.

Based on these models, a capture drainage system was designed, including boreholes down to the alluvial layer located near the tailings repository dam (option 1) and over the area of the alluvial layer (option 2). According to preliminary assessments, this technology is the most effective and the least costly.

Under option 1, approximately 900 m^3 per day (about 330,000 m^3 per year) would be pumped out and transferred to an existing deep repository (storage site) for isolation in a collection layer located at a depth of 1,435-1,600 m. The total volume of industrial wastewater removed from operational production facilities and of water pumped out of the underground layer being rehabilitated will not exceed the designed processing capacity of the storage site, 2,500 m^3 per hour.

Under option 2, which is the more effective, the volume of water pumped out is 2-2.5 times greater, which exceeds the designed processing capacity of the existing storage site. To reduce the volume of water being transferred into storage, plans call for using an ENERGO-70,-45–type reverse osmosis unit. Using such units makes it possible to obtain purified water meeting household consumption or technical use standards for reuse, as well as concentrated solutions (which make up 35 percent of the purified water), which will be transferred to the tailings repository and then pumped through boreholes into the collector layer.

Design work is under way on a technological system for rehabilitating the water-bearing layer according to option 1.

A similar system for cleaning near-surface groundwater layers may be considered for the Mayak enterprise. To bury the water pumped out, which is contaminated with radioactive substances, deep underground collector layers in the Teche-Brodskaya structure (1,000-1,500 m deep) may be used. This structure is located very close to the enterprise and lies partially within the boundaries of its protected sanitary zone.

According to the results of geological prospecting work done in 1960-1965, the Teche-Brodskaya structure was deemed to be unsuitable for the burial of liquid wastes. However, an analysis in recent years of the geological conditions and results of experimental work, taking into account the results of studies of existing liquid radioactive waste storage sites at the Mining-Chemical Complex (Krasnoyarsk Krai), the Siberian Chemical Complex (Tomsk Oblast), and the Scientific Research Institute of Nuclear Reactors (Ulyanovsk Oblast), indicated that the burial of liquid radioactive wastes in the collector layer of the Teche-Brodskaya structure was entirely feasible.

To reduce the volume of water that would be transferred into storage, reverse osmosis desalination units (like the ENERGO-70,-45, for example) and sorption units to extract radionuclides could be used. If this is done, the salt solutions from the reverse osmosis units and the regenerates from the sorption units, which contain radionuclides, would be sent to storage.

To develop the design for the deep repository (storage site) for radioactive water pumped out during rehabilitation of near-surface underground layers, geophysical studies (seismological prospecting) must be carried out to select locations for the drilling of boreholes and deep (up to 2,000 m) exploratory-operational wells in the Argayash Fault zone, very close to reservoirs 10 and 11 of the Mayak enterprise. Studies of the boreholes must also be conducted. If the results are positive, the collector layer between the depths of 1,000 and 1,500 m could also be used to store the liquid radioactive wastes that are currently being transferred into open surface repositories as the reservoirs of the Techa Cascade are being decommissioned and treated.

14

Observations Concerning Mayak

Frank L. Parker, Vanderbilt University

The PowerPoint presentation and the papers about radioactive conditions around the Mayak site in this workshop give us an update on conditions at that site, likely the most radioactively contaminated area in the world, and on possible further remediation efforts.[1] The authors also furnish us with more details than were previously available on the evolution of the dam at Reservoir 11 on the Techa River and on other measures being taken to reduce pressure on the dam and control seepage through it. My colleagues and I have previously explored the possibilities of the collapse of the dams along the Techa River and the radiological consequences that this might entail.[2] It was reassuring to see that measures had been taken to strengthen the dam and relieve the pressures on its face. The impact

[1]Glagolenko, Yu. V., Ye. G. Drozhko, and S. I. Rovny. 2009. Experience in rehabilitating contaminated land and bodies of water around the Mayak Production Association. Pp. 81-91 in Cleaning Up Sites Contaminated with Radioactive Materials: International Workshop Proceedings. Washington, D.C.: The National Academies Press.

Skidanov, V. G., Ye. N. Kamnev, and A. I. Rybalchenko. 2009. Rehabilitation of contaminated groundwater layers near the Mayak enterprise using deep burial technology. Pp. 92-94 in Cleaning Up Sites Contaminated with Radioactive Materials: International Workshop Proceedings. Washington, D.C.: The National Academies Press.

[2]International Institute for Applied Systems Analysis (IIASA). September 1996. Mayak Case Study: Draft Final Report to Lawrence Berkeley Laboratory (DE-AC03-76SF00098-DOE).

of the dam failures had not been published at the time that my colleagues and I made our analyses. However, in 1998, a Russian publication on the distribution of radionuclides on the floodplains of the Techa River was released.[3] These data could have been utilized to validate or reject the results of the modeling and the projected doses to people utilizing the floodplain area.

One of the suggestions in the International Institute for Applied Systems Analysis report, based on some Russian work, was to divert the Techa River upstream of Mayak to the Karabolka River, thereby reducing the amount of water flowing through the Techa River system. No mention is made of this diversion, although there have been stories of such construction. The authors also update us on the filling in of Karachai Lake. The work is 95 percent complete. The discussion of its present impact states that the "forecasts of how the situation may develop over a fairly long period (300 years) indicate that in the future there will be practically no radiologically significant discharge of contaminated groundwater into the open hydrographic network."[4] There is no mention of the closure of water wells in the vicinity that had been an imminent threat earlier nor the likely effect after 300 years.

The environmental discharges from the Mayak enterprise are often compared with those from the Hanford site in the United States. Both were the first production sites for plutonium in their respective countries. However, the resemblance ends there. Though the number of becquerels (curies) discharged to the respective rivers—Techa and Columbia—are similar, the Hanford wastes were almost entirely short-lived induced radioactive nuclides, while the Mayak wastes also consisted of many long-lived fission products. Further, the flow in the Columbia River at that point averages 3,500 m^3 per second[5] and in the Techa River at its mouth the flow was 7 m^3 per second. Consequently, with much lower releases of long-lived radionuclides and much greater dilution, the effects of liquid radioactive releases to the environment have been much lower in the Hanford region than they have been in the Mayak region.[6] For example, no people were displaced from their homes on the Colombia River, while more than 8,000 were moved from the Techa River sites.

The impact of the releases of iodine-131 to the atmosphere from the Hanford

---

[3]Govorun, A. P., A. V. Chenokov, and S. B. Shcherbak. 1998. Distribution of ^{131}Cs inventory in the floodplain of the Techa River in the Muslyumovo village region. Atomic Energy 84(6).

[4]Glagolenko et al., op. cit.

[5]Evans, R. G., M. J. Hattendorf, and C. T. Kincaid. February 2000. Evaluation of the Potential for Agricultural Development at the Hanford Site, PNNL-13125. Available online at *www.osti.gov/bridge/servlets/purl/751663-dcRc24/webviewable/751663.PDF*.

[6]Akleyev, A. V., and M. F. Kisselyov, scientific editors (translators K. M. Zhidkova and K. A. Akleyeva). 2002. Medical-Biological and Ecological Impacts of Radioactive Contamination of the Techa River. Chelyabinsk: Fregat.

Health Physics: The Radiation Safety Journal 93(3), September 2007. The entire issue is devoted to radiological conditions at the Mayak site.

site has been extensively studied.[7] The impact on "downwinders" near the Hanford site was greatest for children who received an average thyroid dose of 235 rads, and at the most impacted areas the doses ranged from 54 to 870 rads.[8] Until now, no such studies of the iodine-131 releases from Mayak have been published, but are due to be published in 2008.[9]

The paper by Skidanov[10] and a companion paper by Rybalchenko[11] caused the most comment at the meeting because of the change in view on the viability of deep geological disposal near the Mayak site. This brought a response that excessive amounts of tritium have been found in the vicinity of the deep well injection system at Krasnoyarsk. Rybalchenko responded that the tritium was due to surface operations at the plant. However, Nosov et al. wrote, "The tritium concentration in the Podporogovy stream . . . is an indicator of the possible relation between surface waters and the region of unloading of the underground levels, which are collectors for pumping liquid radioactive wastes on the Severny test area."[12] In addition, Bolsunovsky and Bondareva state, "In water and sediment samples of the Bolshaya Tel River (a tributary of the Yenisei River) the tritium content turned out to be at least 10 times higher than background values of the Yenisei River. This allows the conclusion that there is water exchange between the surface waters and the radioactively contaminated underground horizons of the Severny site."[13] Finally, Kasyanova states, "In our country, since the 1960s, radioactive waste have been stored underground in the regions of Tomsk, Krasnoyarsk, and Dimitrovgrad . . . . The risk associated with these objects is high. Extreme accident situations due to caused underestimation of the characteristic features of the spatiotemporal changes in the development of present-day geodynamic processes have not been ruled out here."[14] It appears that the only way to settle this argument is to actually sample the projected flow paths to determine if there are greater concentrations of tritium there than in background samples.

---

[7]The Technical Steering Panel of the Hanford Environmental Dose Reconstruction Project. 1994. Summary: Radiation Dose Estimates from Hanford Radioactive Material Releases to the Air and the Columbia River. Washington State Department of Ecology.

[8]Gephart, R. E. 2003. Hanford: A Conversation about Nuclear Waste Cleanup. Columbus, OH: Battelle Press.

[9]Glagolenko, Yu. V., Ye. G. Drozhko, Yu. G. Mokrov, N. P. Piatin, S. I. Rovny, L. R. Anspaugh, and B. A. Napier. In press. Method and results of reconstruction of radioactive noble gas releases from graphite reactor stacks of Mayak Production Association for the total period of its operation. Radiation Safety Problems (Mayak Production Association Scientific Journal).

[10]Skidanov et al., op. cit.

[11]Skidanov et al., op. cit.

[12]Nosov, A. V., A. M. Martynova, V. F. Shabanov, Yu. V. Savitskii, A. E. Shishlov, and Yu. A. Revenko. 2001. Investigation of the tritium transport by water flows from the territory of the Mining-Chemical Combine in Krasnoyarsk. Atomic Energy 90(1).

[13]Bolsunovsky, A. Yu., and L. G. Bondareva. 2003. Tritium in surface waters of the Yenisei Basin. Journal of Environmental Radioactivity 66.

[14]Kasyanova, N. A. 2002. Safety of deep burial of radioactive wastes. Atomic Energy 93(1).

Until that is done, there will continue to be uncertainty about the safety of the deep injection disposal sites in Russia.

It is unfortunate that the papers presented on Mayak contained no references, so more detailed information was not easily available.

I am indebted to my Russian and American colleagues for more detailed discussions on these topics and the detailed discussions on remediation of Russian sites covered in Alexakhin et al.[15]

[15]Alexakhin, R. M., L. A. Buldakov, V. A. Gubanov, Ye. G. Drozhko, L. A. Ilyin, I. I. Kryshev, I. I. Linge, G. N. Romanov, M. N. Savkin, M. M. Saurov, F. A. Tikhomirov, and Yu. B. Kholina. 2004. Large Radiation Accidents: Consequences and Protective Countermeasures. Moscow: IzdAT.

15

Remediation of Contaminated Facilities at the Kurchatov Institute

V. G. Volkov, Yu. A. Zverkov, S. G. Semenov, A. V. Chesnokov, and A. D. Shisha, Russian Research Center—Kurchatov Institute

INTRODUCTION

During many years of research activities to develop nuclear technologies for military and civil applications, the Russian Research Center—Kurchatov Institute accumulated at its site considerable amounts of solid radioactive waste and spent nuclear fuel.[1] Solid radioactive waste produced before the mid-1960s, some of it having high specific activity, had been placed in interim storage at a special site within the Kurchatov Institute. According to initial estimates, 1,200 m^3 of radioactive waste with a total activity of about 3.7 × 1015 Bq (100,000 Ci) at the time of disposal were placed in temporary storage facilities at the site. At present, because of intensive construction in Moscow, the institute is surrounded by a densely populated district of the city, and the radioactive waste disposal site adjoins an urban residential area.

[1]Ponomarev-Stepnoi, N. N., V. G. Volkov, N. Ye. Kukharkin, et al. 2002. Rehabilitation of radioactively contaminated facilities and the site of the Russian Research Center—Kurchatov Institute. Conference Handbook of IBC's Eighth International Conference and Exhibition on Decommissioning of Nuclear Facilities—Managing the Legacy, London, United Kingdom, November 11-12, 2002, EA1141, IBC Global Conferences.

MIGRATION OF RADIOACTIVE WASTE CONTAMINATION INTO GROUNDWATER

To evaluate the environmental impact of old radioactive waste sites and to predict the spread of radioactive contamination through groundwater, monitoring boreholes were drilled at the waste disposal site and areas adjoining it on the south and west. Equipped with filter columns, these boreholes were designed to permit observation of the level, chemical composition, and radionuclide content of the groundwater.

Available data on the geological structure of the soil, permeability coefficients of subsurface horizons, groundwater level, and volume of radioactivity served as a basis for calculation of the groundwater flow structure and strontium-90 dispersal range.[2] Apparently due to a sharp increase in the groundwater level in the early 1990s, the groundwater flow structure changed, resulting in a risk of contamination beyond the Kurchatov Institute. In addition, according to a model created to predict strontium-90 dispersion, if remediation work is done on the temporary radioactive waste sites, the area of groundwater contamination with strontium-90 content exceeding the action level (5 Bq/L) will remain within the Kurchatov Institute buffer zone.[3] If remediation activities are not performed, the contamination may spread further with groundwater beyond the disposal site, with the strontium-90 content exceeding the action level.

PROPOSED MEANS OF REMEDIATION OF THE RADIOACTIVE WASTE DISPOSAL SITE

Initially, two alternative approaches to remediation of the radioactive waste disposal site were considered:

1. Creation of geophysically engineered barriers in the radionuclide migration pathways
2. Complete disposition of the waste sites and decontamination of radioactive soil

To estimate the cost of the first option, evaluations were made of the possibility of building engineered barriers to reduce substantially the spread of contamination by infiltration with simultaneous use of sorbents to extract strontium-90 from the groundwater. Zeolites and apatites were considered as sorbents, and underground leaching technologies were proposed to extract cesium-137 from soils.

---

[2]Rastorguev, A. V., K. Bukharin, V. G. Volkov, et al. 2005. Prognosis of radionuclide contamination spreading on the site of temporary waste storage of RRC Kurchatov Institute. Proceedings of the International Congress ECORAD 2004: The Scientific Basis for Environment Protection Against Radioactivity, Aix-en-Provence, France, September 8-10, 2004. Radioprotection 40(Supp.1):367-370.

[3]Op. cit.

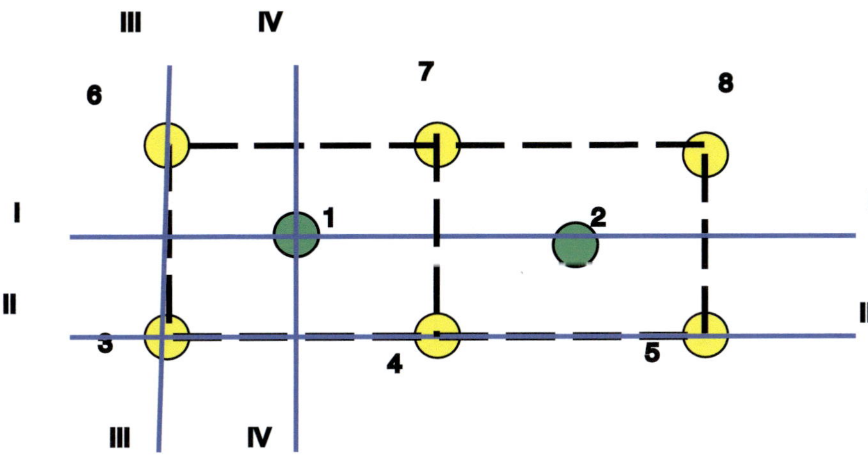

FIGURE 15-1 Basic layout of boreholes for the "double envelope" design.
NOTE: Boreholes 1 and 2 are for exhaust (optimum flow rate $Q_{1,2} = 0.3\ m^3/d$), and boreholes 3-8 are for injection (optimum flow rate $Q_{3\text{-}8} = 0.1\ m^3/d$).

These steps would involve a large volume of boring operations and high material costs. In the meantime, to evaluate the efficiency of such countermeasures, specialists from the Radon enterprise performed laboratory studies on conditions that would permit efficient chemical reactions involved in underground leaching.[4] The studies were performed using near-bottom soils from Temporary Storage Facility No. 3 (the contaminated area was $10 \times 8\ m^2$ and 3 m thick) as an example. On the basis of geological and hydrological information and information on the morphology of the area contaminated with cesium-137 within Storage Facility No. 3 at the radioactive waste disposal site, an optimum design for a "double envelope" underground leaching approach involving two exhaust and six injection process boreholes was selected. The basic layout of the boreholes is shown in Figure 15-1.

Based on laboratory studies, a mixture of sulphuric and phosphoric acids (0.25M H_2SO_4 + 0.25M H_3PO_4) was selected as a process solution. The process solution was injected through boreholes 3-8 and pumped out through boreholes 1-2. The optimum flow rate values for this scheme were estimated as follows: for exhaust boreholes, $Qe = 0.3\ m^3/d$, and for injection boreholes, $Qi = 0.1\ m^3/d$. Because of clogging, this design resulted in flows that were tens of times less than the flow under ideal conditions, when the soil contains no organic inclusions. Calculations were then made to estimate the time required to clean up the near-bottom area of the cesium-137–contaminated Storage Facility No. 3. They totaled t = 4.55 yr at efficiency E = 84.6 percent and t = 2.14 yr at E = 60 percent

[4]Lunev, L. I. 1982. Application Conditions and Physicochemical Basics of Underground Uranium Leaching. Moscow: MGRI Publishing House.

(attainment of the residual specific activity of sand $C_t = 10$ Bq/g). The foregoing calculations assumed that it would be possible to solve the problem of carbon dioxide release associated with the leaching process. If there is no acceptable technical option for solving this problem, the above calculations will have no practical use.

In conclusion, the studies demonstrated that cleaning cesium-137 from rock enclosures under conditions prevalent at Facility No. 3 at the radioactive waste disposal site presents major technical and technological difficulties. Therefore, underground leaching cannot yet be proposed for immediate implementation. According to estimates, the work required to stabilize the waste sites entails high material costs. Therefore, it was decided to pursue the second option, which involves complete disposition of the sites. After remediation, the radioactive waste disposal site will still belong to the Kurchatov Institute, so standards established for the institute and its personnel were accepted as the remediation standard. Table 15-1 presents reference levels for background and soil contamination at the disposal site and technogenic gamma background dose rates (above the natural background). These levels are officially established by Kurchatov Institute Order No. 74, dated February 29, 2000, and by sanitary-epidemiological findings of the Russian Federation Ministry of Health Agency for Sanitary-Epidemiological Inspection. As specified by the groundwater remediation criterion, the action level for possible population exposure was 5 Bq/L for strontium-90.

TABLE 15-1 Reference Levels of Technogenic Background and Soil Contamination at the Radioactive Waste Disposal Site

Monitored Parameter	Value	Note
Equivalent technogenic gamma background dose rate at the radioactive waste disposal site	2.5 μSv/hr (250 μrem/hr)	Mean integral dose at Kurchatov Institute perimeter: 0.8 mSv/yr (in 2005)
Gamma background on the soil surface at the disposal site	up to 3.0 μSv/hr (up to 300 μR/hr)	Soil may be used for filling pits, holes, trenches, and so forth at disposal site
	above 3.0 μSv/hr (above 300 μR/hr)	Soil must be decontaminated or removed for disposal

TECHNOLOGIES FOR DISPOSITION OF THE OLD WASTE SITES

Given the lack of accurate data on design features of the old sites and composition of the radioactive waste they held, disposition of the sites was performed in accordance with the following standard sequence of steps:[5]

[5]Volkov, V. G., G. G. Gorodetsky, Yu. A. Zverkov, et al. 2004. Radwaste management technologies used in remediation of radioactively contaminated facilities and areas of RRC Kurchatov Institute. Pp. 141-156 in Proceedings of the Seventh International Conference "Nuclear Technology Safety: RW Management," September 27-October 1, 2004, St. Petersburg, Russia. St. Petersburg: ProAtom Publishing House.

- Drilling of exploratory boreholes in the repository boundary areas and radioactive waste mass followed by a radiation survey
- Removal of the built-up ground from the storage facilities and destruction and removal of the facility roofs
- Extraction of waste from the facilities, waste sorting, and waste loading into certified containers
- Inspection and disposition of storage facility structural elements
- Sorting of soil and removal of contaminated soil from storage facility pits
- Final radiation survey of storage facility pits and their backfilling with clean soil

During disposition of the sites containing concrete-encased high-level waste, shadow radiation shielding was built around them. Exploratory boreholes were made in the site boundary areas and radioactive waste mass to determine more precisely the locations of the repositories, their geometric sizes, and their design features, as well as to perform radiation surveys. Equivalent gamma dose rates were measured along the depth of the boreholes using a certified UIM-2-2 instrument with a BDMG-100 detector. Collimated detectors were used to measure distribution of specific activities of gamma-emitting radionuclides. Visual inspection of the boreholes was performed with a specially developed compact video camera, and the signal was recorded on a computer.

Built-up ground was removed from the facility roofs and roof openings using conventional construction machines equipped with the necessary attachments, depending on the type of work. A truck crane was used to lift easily removable roof slabs. Cast–in situ concrete roofs were destroyed using EK-12 and EK-270 excavators equipped with hydraulic hammers. In individual cases where a roof was found to be made of thick monolithic concrete (for example, during disposition of Repository No. 2), the roof was destroyed with a device for electric-discharge demolition of concrete structures.[6] This technique uses electric-discharge energy released in condensed media in a plasma channel as a high-power current pulse passes through the channel. To accomplish this, shot holes were drilled in the monolithic concrete and filled with water, after which an electric discharge was produced that broke the in situ concrete slab into fragments. In addition to this remediation work, dust-suppression techniques were applied and aerosol activity in the atmosphere in the working area was monitored.

The presence of high-level radioactive waste in Repository No. 4 required construction of additional radiation shielding around it. Radiation calculations were performed for several shielding designs, taking into account the particular characteristics of the repository location, geometrical parameters of the space to

[6]Smirnov, V. P., Ye. G. Krastelev, V. M. Nistratov, et al. 1999. Development and application of a mobile facility for electro-discharge destruction of rock and structures. Mining Journal 11:56-58.

be opened, and approximate ratios between specific activities of the major dose-producing radionuclides, cobalt-60 and cesium-137.

For operations at Storage Facility No. 4, radiation shielding was designed and constructed based on the calculation results. The roof consisted of 6-m-long, 20-cm-thick paving slabs resting on outer support walls built of foundation blocks measuring 2,400 × 600 × 400 mm reinforced with metal trusses. A labyrinth was built to allow robots to pass into the shielding while preventing ionizing radiation from streaming out beyond the shielded area.

The radioactive waste was extracted from the old facilities using conventional wheeled and crawler construction machines as well as Swedish Brokk-110 and Brokk-330 robots. Low-level waste was extracted using front loaders. The robots removed intermediate-level waste and fragments of high-level waste. To protect operators against ionizing radiation, construction equipment cabs were shielded with lead sheets and provided with protective leaded glass. Both construction equipment and robots were equipped with collimated detectors to measure the activity of the radioactive waste being extracted.

During operations at the high-level waste storage facility, color video monitoring cameras were installed inside the radiation-shielding structure. Their signals were received by monitors located in the excavator cabs. To warn personnel about radiation hazards, working areas were equipped with threshold collimated detectors that produced audible and light alarms when the allowed gamma dose rate level was exceeded.

The concrete matrix of the facility was gradually destroyed through the opening in the radiation shielding by an excavator with a hydraulic hammer located on the shielding roof. Fragments of the broken waste mass that contained low- and intermediate-level waste and concrete spalls were removed from under the shielding structure by a front loader and placed into metal or reinforced concrete containers. The container type to be used for loading the waste was selected based on measurements taken by a collimated detector mounted on the top edge of the shielding structure.

To detect canisters of high-level waste or their fragments in the destroyed concrete mass, operators used a gamma camera that transmitted its signal to a monitor.[7] When canisters or high-level waste fragments were detected, the shielding roof was put in place completely, and robots then extracted, measured, and packaged the high-level waste inside the shielding structure without any personnel being exposed to high radiation fields. The robots destroyed the remains of the waste concrete matrix, gripped the detected high-level waste, and removed it

[7]Volkov, V. G., A. G. Volkovich, A. S. Danilovich, O. P. Ivanov, S. V. Smirnov, and V. Ye. Stepanov. 2005. Application of new instruments for radioactive waste sorting in remediation activities at RRC Kurchatov Institute. Pp. 135-141 in Proceedings of the Eighth International Conference "Nuclear Technology Safety: Economy and Management of Ionizing Radiation Sources," September 26-30, 2005, St. Petersburg, Russia.

to a special sorting area arranged inside the shielding structure. Inside the shadow shielding, the robots performed all operations required to extract and cut up the waste and then pack it into containers. The gamma camera display was used to direct the robots to high-level radiation sources and monitor the extraction, fragmentation, and packaging operations. High-level waste was removed from under the shadow shielding and packed in metal containers with concrete inserts that were then grouted in concrete at the top. Before shipping the filled containers to the Radon enterprise, gamma levels near the containers were checked to ensure that they met regulatory requirements.

DISPOSITION OF THE OLD REPOSITORIES

All radioactive waste management operations involved continuous radiation monitoring, including measurement of activity levels of the waste loaded into containers, as well as continuous monitoring of working areas, personnel, and the volume of activity of radionuclides in the working area. Activity levels of radioactive waste in containers were measured with spectrometric and current-collimated detectors. Measurement results were further processed with custom-developed software, taking into account the container geometry, thickness, and material; the waste-packaging density; and the ratio between activities of the basic radionuclides (cobalt-60 and cesium-137) found in the radioactive waste.[8] The measurements were made according to approved procedures using certified collimated detectors to monitor specific activity.

All dosimetry instruments used for personal and overall radiation monitoring had been registered and entered in the Register of Instruments. For spectrometric and radiometric measurements of waste, soil, groundwater, and air samples performed in laboratory conditions, certified instruments and qualified procedures were used. In addition to working areas, the radiation situation was monitored at the entire radioactive waste disposal site. During operations on Storage Facility No. 4, the radiation situation in working areas and at the entire disposal site was monitored with two gamma locators that measured the ionizing radiation photon flux, taking into account its spectral characteristics.[9]

One of the gamma locators was used for continuous monitoring of changes

---

[8]Volkov, V. G., V. N. Potapov, O. P. Ivanov, S. M. Ignatov, N. K. Kononov, and V. Ye. Stepanov. New radiation monitoring instruments and systems and their use in remediation operations at the RRC Kurchatov Institute radioactive waste disposal site. Pp. 371-378 in Proceedings of the Seventh International Conference "Nuclear Technology Safety: RW Management," September 27-October 1, 2004, St. Petersburg, Russia. St. Petersburg: ProAtom Publishing House.

[9]Volkov, V. G., A. G. Volkovich, A. S. Danilovich, O. P. Ivanov, S. V. Smirnov, and V. Ye. Stepanov. 2005. Application of new instruments for radioactive waste sorting in remediation activities at RRC Kurchatov Institute. Pp. 135-141 in Proceedings of the Eighth International Conference "Nuclear Technology Safety: Economy and Management of Ionizing Radiation Sources," September 26-30, 2005, St. Petersburg, Russia.

in the radiation situation in working areas at Storage Facility No. 4, with the measurements displayed on a computer screen via the Internet. The other gamma locator scanned the entire disposal site, measuring gamma spectra from individual areas. Photon flux values measured at the gamma locator position were normalized to the gamma exposure dose rate (EDR) value measured by an integral dosimeter mounted on a rotator of one of the gamma locators. Photon flux distributions obtained were further used to calculate gamma EDR values in all points of the scanned space, with these data presented as a color palette superimposed on a coordinate image of the scanned object.

During remediation operations, the volume activity of aerosols in the working area air was monitored during each shift using UDA-1AB devices and PVP-4A samplers and at the site perimeter 24 hours a day using Typhoon-type devices. These devices continuously measured the volume of activity of alpha- and beta-active aerosols in the air throughout the operations. During the measurements, the systems were typically located around the working areas about 3-5 m apart, with air samples taken at the height of about 1 m above the ground.

The samplers independently monitored the level of air contamination by radioactive aerosol mixtures of various compositions. Air samples were taken while pumping a controlled volume of air through a special filter that was subsequently delivered for spectrometric and radiochemical analysis under laboratory conditions after the sample was taken. The samplers were equipped with a self-contained power supply that allowed air samples to be taken in remote, hard-to-reach locations where the main power supply is inaccessible. The control interval and air sampling points were determined depending on the kind of work being performed, locations of dust-producing activity sources in working areas, and personnel workplaces. The samplers were placed about 1 m from the dust-generating activity sources at an elevation of about 0.5 m above ground level. During operations on Storage Facility No. 4, the samplers were installed not only at personnel locations but also at locations of robotic operations to further evaluate the dust rise and dust-suppression efficiency directly in working areas. In addition to monitoring the overall radiation situation, site managers also monitored groundwater activity using a network of observation boreholes located at the radioactive waste disposal site and in the adjoining area.

APPLICATION OF DUST-SUPPRESSION TECHNOLOGIES

To ensure radiation safety for personnel involved in the work and prevent formation and transfer of radioactive aerosols, all operations involving radioactive waste extraction and disposition of the old storage facilities were performed using dust-suppression technologies. Various localizing, isolating, and dust-suppressing protective coatings based on polymeric compounds were used in the operations. These polymeric coatings were capable of preventing the spread of radioactive contamination in the form of dispersion aerosols into the environment. The protective polymeric coatings were applied to dust-prone surfaces using airless and airstream spraying techniques. The efficiency of using these compounds was controlled by measuring (1) the surface activity of radionuclides in smears taken from the external surface of the coatings applied to the surfaces to be protected and (2) the volume activity of aerosols in the working area air.

DECONTAMINATION OF RADIOACTIVE SOIL AND METAL RADIOACTIVE WASTE

Disposition of the radioactive waste repositories requires solving problems of how to decontaminate large volumes of radioactive soil and establish a system for post-remediation monitoring of the areas involved. Estimates of radioactive soil volumes suggested that high-performance (2-3 metric tons per hour) soil-cleaning technologies were required. The search for such technologies, which reviewed chemical, electrokinetic, and other soil decontamination methods, indicated that they would have to be developed during the remediation operations.[10] Two radioactive soil treatment technologies were selected based on performance estimates: dry radiometric separation and water-gravity separation.[11] Further efforts were devoted to develop these two technologies and bring the performance of the facilities built around these technologies up to required levels.

According to the results of preliminary studies performed by the Bochvar Institute, more than 80 percent of radionuclides in the contaminated soil are accumulated in fine sludge or clay fractions or both. Therefore, the technology of wet decontamination of soil consisting of water-gravity separation and classification of the contaminated soil into size fractions followed by segregation and removal of the fine fraction was adopted for soil decontamination. Based on laboratory findings, the optimum approach is to separate contaminated soil into the following four fractions by size: fraction 1 (lump), greater than 100 mm in size; fraction 2 (coarse), from 3 to 100 mm; fraction 3 (sand), from 0.1 to 3 mm; and fraction 4 (fine or sludge), up to 0.1 mm.

A pilot facility for wet decontamination of soil was developed around this technology in cooperation with the Bochvar Institute and the Chemical Technology Institute. The basic equipment for this pilot facility was fabricated at the Gormashexport enterprise in Novosibirsk. It has a modular design consisting of the three separate components for disintegration, classification, and thickening. This pilot facility for water-gravity separation of contaminated soil was installed at the radioactive waste disposal site. Operation of the pilot facility in

[10]Dmitriev, S. A., L. B. Prozorov, M. Yu. Shcheglov, et al. 2001. Electrokinetic method of cleaning radionuclides from soil. Radiation Safety Issues 1:42-49.

[11]Volkov, V. G., Yu. A. Zverkov, S. M. Koltyshev, et al. 2005. Main results of start-up and trial operation of the soil decontamination facility. Pp. 120-135 in Proceedings of the Eighth International Conference "Nuclear Technology Safety: Economy and Management of Ionizing Radiation Sources," September 26-30, 2005, St. Petersburg, Russia.

the start-up and adjustment modes has demonstrated its rather high efficiency. The specific activity of 70-80 percent of the initial soil is reduced by four to five times. On average, from 180 to 200 kg of each metric ton of initial soil processed are removed for long-term storage. Recovered water used in processing remains virtually uncontaminated throughout several facility operation cycles. At present, more than 2,000 m^3 of radioactive soil have been processed and about 400 m^3 have been shipped to the Radon enterprise for long-term storage.

Of the radioactive waste extracted from the repositories, up to 30 percent is composed of contaminated metal and metal structures, with these materials primarily classified as low-level waste. Hydroabrasive technology is used to decontaminate the metal waste. A hydroabrasive cutting and decontamination facility was deployed in a specially equipped area. The objects to be processed are placed on reinforced concrete plates resting directly on the metal floor. The hydroabrasive decontamination equipment consists of a high-pressure pump, high-pressure piping, a bin, and a feeder supplying loose abrasive to a hydroabrasive nozzle. The nozzle may be used either as a cutter or as a sprayer to remove an upper contaminated layer from metal surfaces.

The decontamination is performed both manually and using automated tools. In the manual mode, cleaning is accomplished using an operating water pressure of 1,500 atm at an abrasive sand flow rate of about 1 kg per minute. In the automated mode, decontamination is performed by scanning metal surfaces with a water-abrasive jet with a linear velocity of 500 mm per minute using a remotely controlled cart. The operating water pressure used for cleaning is 2,000 atm with the nozzle diameter being 0.6 mm.

The hydroabrasive equipment demonstrated good efficiency for low-level metal waste decontamination. However, as the level of surface contamination increases, decontamination costs increase because of higher consumption of abrasive material and, consequently, increased quantities of secondary waste produced in the decontamination process (abrasive, wastewater, and so forth). Cost estimates allowed the most cost-effective decontamination levels to be selected in order to make the decontamination process economically sound. About 250 metric tons of metal radioactive waste was decontaminated using the hydroabrasive equipment and then shipped to the Ecomet-S enterprise for remelting.

MAIN RESULTS OF THE WORK

All 10 old radioactive waste repositories subject to remediation at the waste disposal site have been decommissioned and eliminated. About 3,400 m^3 of radioactive waste with a total activity of more than 1.4×10^{13} Bq (approximately 380 Ci) was extracted from the repositories, and more than 3,000 m^3 was moved to Radon for long-term storage.

During 2005 the gamma dose rate at the radioactive waste disposal site perimeter varied between 0.21 and 0.67 μSv per hour and at the institute perimeter,

TABLE 15-2 Personnel Exposure During Remediation Operations (2003-2006)

Parameter	2003	2004	2005	2006
Mean individual dose, mSv	2.0	1.44	1.8	1.9
Collective dose, Man × Sv	0.041	0.080	0.090	0.079

between 0.08 and 0.14 μSv per hour. The volume of activity in the air at the disposal site and within the overall Kurchatov Institute site was by three to five orders of magnitude less than allowable levels for urban populations. External radiation exposure of personnel by years is presented in Table 15-2.

Thus, the effective organization of work, the equipment used for operations, and radiation-monitoring instruments allowed this repository containing high-level and other waste to be decommissioned and eliminated rather quickly while meeting radiation safety requirements for onsite personnel and the local population within a major city. Results of the 3-year remediation effort have demonstrated that the choice of low-, intermediate-, and high-level waste management technologies was correct. During 2005, Repository No. 4, where canisters of high-level waste were encased in strong concrete, was successfully decommissioned. Organizational and technical measures that were applied made it possible to complete the work in the shortest possible time in full compliance with all rules

16

Selected Remediation Issues at the Russian Research Center—Kurchatov Institute

Roy E. Gephart, Environmental Molecular Sciences Laboratory, Pacific Northwest National Laboratory

The Russian Research Center—Kurchatov Institute (RRC—KI) is the leading institute in the former Soviet Union devoted to military and civilian nuclear programs. Founded in 1943 in the outskirts of Moscow, this 100-ha site of nearly undeveloped, prime real estate is now surrounded by densely populated urban and business districts. Some public housing adjoins the site's outer perimeter. Today there are growing concerns over the public safety and environmental security of the site resulting from increasingly obsolete nuclear facilities and a legacy of inadequate waste management practices that resulted in contaminant releases and challenging remediation problems. In addition, there is growing worry over the presence of nuclear facilities within urban areas creating potential targets for terrorist attacks.

During our visit to RRC—KI, officials shared that the useful lifetime for all onsite reactors is nearly complete, and the institute is working toward decommissioning those reactors and removing all spent fuel by 2015. These efforts will be coordinated with work to complete another facility and carry out environmental remediation activities. Cleanup schedules are dependent on funding.

Based on meetings with RRC—KI staff and reading information about the history and remediation of contamination at the site, the following comments are offered.

Site Inventory of Nuclear Material and Waste. The master plan for remediation of RRC—KI should contain an inventory of the nuclear material and waste remaining onsite. This permits decision makers to understand potential radiological and chemical risks as well as changes in risk profiles and dose rates resulting from remediation work. Written records, interviews with former workers, extrapolations, and new targeted site investigations form the basis for these inventories. Constructing such an inventory is challenging because of incomplete and sometimes conflicting data. The U.S. Department of Energy has faced similar challenges at its nuclear material production, weapons manufacturing, and testing sites.

This inventory could include, for example, information quantifying damaged vs. undamaged experimental spent fuel elements, buried waste, soil and groundwater contamination, surface facility hazards, and orphaned radioactive sources or scattered contaminated spots. Such knowledge assists decision makers in building factual cases for funding site remediation programs. According to the Government of Moscow's Resolution No. 641-PP, On Accelerated Decommissioning of Radiation Hazard Facilities at RRC—KI, the goal of rehabilitation activities is to "eliminate all potentially hazardous sources of radiation that may produce adverse environmental effects, and to transform RRC—KI into a secure and safe nuclear research center within the Moscow city limits" (Volkov et al., 2003). Maintaining an up-to-date material and waste inventory is essential for achieving this resolution.

Underground Water Pipes and Drainage Systems. Rastorguev et al. (2005) spoke of a groundwater level rise averaging 3 m plus changes in groundwater flow directions and peak strontium-90 activities in the shallow aquifer beneath RRC—KI taking place between the late 1980s and the early 1990s. The report continued by stating that these changes were likely attributed to "leakage from the city sewer that crosses the radwaste disposal site," resulting in the partial submergence of some buried waste sites and flushing out of contamination. Volkov et al. (2003) wrote that the institute's sewerage system has "undergone no repairs" since construction, surveys have uncovered ruptured pipelines, and the city's sewerage system, apparently crossing the site, is a potential "source of adverse environmental impacts" that could cause heavy flooding of the site. These are serious concerns in efforts to minimizing subsurface contaminant migration off the RRC—KI site.

Consideration should be given to testing the structural integrity of key water and waste pipelines, abandoning and grouting those of questionable integrity or those of "defunct branches" (Volkov et al., 2003), and installing new lines as necessary. Attention should also be given to RRC—KI installing its own water drainage system to intercept, control, and treat (if necessary) water runoff, especially during torrential downpours when the potential for resuspension of surface contamination is greatest. This would address one of the major uncertainties in the modeling of onsite hydrologic conditions and estimating offsite radiation doses through lessening a major source of contaminant spread as well as potential changes to groundwater flow patterns and rates. These recommendations require the application of standard engineering practices common to municipalities and industries.

Contaminant Barriers. Volkov et al. (2009) reported that the use of zeolites and apatites in permeable subsurface barriers to absorb strontium-90 or a mixture of sulphuric and phosphoric acids to leach cesium-137 from underground sediments was too expensive and experimental for application at RRC—KI's underground waste repositories. Nonetheless, continued examination of such innovative technologies or the use of surface-engineered barriers to control water infiltration (and thus radionuclide migration) is encouraged.

Consideration should be given to examining the feasibility of site-tailored surface-engineered barriers such as those installed at the Hanford site in the U.S. state of Washington. Built in 1994, the Hanford barrier covers 2 ha and is constructed from multiple layers of natural sediments and man-made materials to control moisture, plant, and animal entry while minimizing erosion and moisture infiltration even under extreme storm events. Such barriers are nearly maintenance free for hundreds of years and would also control the suspension of contaminated dust.

In addition, a 90-m-long permeable subsurface reactive test barrier using apatite sequestration to inhibit the migration of strontium-90–contaminated groundwater flow into the nearby Columbia River is also being installed at the Hanford site. Such technology might be applicable to controlling the spread of the strontium-90 plume beneath RRC—KI.

Excavation of Waste Repositories. Between 2003 and 2006, 3,400 m^3 of solid radwaste was excavated from 10 old subsurface concrete waste repositories at RRC—KI. Conventional and modified construction equipment was used to access, remove, and repackage waste for onsite disposal or offsite shipment to the Joint Environmental-Technological Scientific Research Center for Radioactive Waste Decontamination and Environmental Protection (MosNPO Radon). Radiation-shielded areas were built for robotic waste retrieval when intermediate to highly radioactive materials were uncovered. Studies reported in Volkov et al. (2009) suggested that rapid removal of these repositories would be more cost effective than constructing engineered barriers and would accelerate the removal of subsurface contamination sources. The recommended remediation approach appeared reasonable, although concerns are raised over some observations of worker safety.

For comparison, I will use a somewhat analogous though nonurbanized example to the RRC—KI radwaste repository removal—the Accelerated Retrieval Project in Pit 4 at the Idaho National Engineering and Environmental Laboratory (INEEL) in the state of Idaho. Pit 4 cleanup involved construction of a tent-like enclosure covering the entire low-level and transuranic waste burial site. This fully enclosed all excavation equipment and workers. Water spray was sometimes used to suppress dust during the warmer months at the RRC—KI site. Otherwise, most remediation was conducted in the open air. All workers inside the Pit 4 enclosure wore fully protective, tape-sealed clothing plus full-face filtered masks. Pictures of the RRC—KI waste removal frequently showed workers without hardhats or particle masks and wearing loose-fitting street-type clothing. While visiting RRC—KI on June 8, 2007, and observing solid low-level waste removal using heavy equipment, onsite workers were lifting pipes and other heavy objects overhead and stirring dust. However, basic safety equipment appeared missing—no hardhats, no particle masks, and minimal dust suppression using periodic water spray. Nearby residences (perhaps 100 m away) went unprotected and perhaps uninformed of cleanup activities.

Volkov et al. (2009) wrote that detection of high-level waste fragments in the solid waste was accomplished using gamma counter-equipped cameras with a signal display on an operator's monitor. Questions arise about the potential for acute worker exposure between the time of fragment detection and use of protective roof shielding and robotics for further material handling. Based upon available information, concerns also exist about the effectiveness of air sampling for alpha- and beta-bearing aerosols suspended around remediation sites when sampling filters are taken to a laboratory for spectrometric and radiochemical analyses before potential worker exposures are recognized, as well as the potential need for increased dust abatement to be implemented.

The collection and treatment of wastewater created when washing trucks removing solid waste and debris from surface excavations before traveling on public roads is applauded.

Groundwater Modeling and Environmental Monitoring. Existing hydrologic models are not based on "very rigorous site-specific features" and are, therefore, thought useful for a "first approximation" and inference of flow and transport estimates (Novikov et al., 2005; Novikov, 2007). Missing information was obtained using computational-analytical methods. There is a need for more site-specific information on such parameters as sediment hydraulic conductivity, water runoff, hydraulic heads, soil moisture, water infiltration, and the physicochemical characteristics of buried waste and subsurface contaminants to validate computational models, reduce modeling uncertainty, and more reliably use modeling results to predict present and future flow system behavior. Knowing the hydraulic properties and distributions of highly reworked, nonuniform shallow soil and rubbish mixtures (e.g., from past building demolitions, sediment excavations, ravine filling) discarded over the years is critical because they could dominate water infiltration within contaminated areas. Novikov (2007) notes the permeability of these deposits "was not studied." Based upon available data, have alternative, though equally valid, flow and transport models been developed?

Consideration should be given to the installation of soil lysimeters for quantifying water infiltration and low-volume (to minimize water extraction) hydraulic tests conducted for measuring sediment permeability. Hydraulic head distribution maps would also be useful to model lateral and vertical flow potentials.

Rastorguev et al. (2005) stated that water level observations in boreholes have been discontinued except inside wells drilled since 2002-2003. Based upon information reported in Novikov (2007), Rastorguev et al. (2005), and Volkov et al. (2003), there appear to be 17 to 30 boreholes used for water sampling. The actual number was unclear. Are water samples drawn from different subsurface horizons to identify the lateral and vertical extent of groundwater contamination and off-site migration? A sustained commitment to long-term environmental monitoring at select sites to establish radiation exposure baselines is necessary to quantify environmental risks and to gain benefits from site remediation efforts.

This writer assumes that periodic groundwater and environmental monitoring reports are published. These would include, for example, contaminant distributions, points of dosimetric monitoring and environmental sampling, plus average and maximum worker and public health effective dose equivalents from exposure to RRC—KI contaminants. It would be useful to report the distribution of environmental risks the public receives from the various pathways—water, air, and food. Such information enables decision makers to focus cleanup efforts where the greatest risk reduction benefits would take place.

Public Involvement. Novikov (2007) addresses the issue of reducing "public anxiety" about radioactive releases from RRC—KI. This concern was also noted in the presentation given by Volkov on June 5, 2007, as well as in other talks discussing remediation progress at contamination sites across Russia. However, few specifics describing stakeholder engagement were provided. Nearly 30 years of experience in the United States implementing federal waste management and cleanup regulations under the Resource Conservation and Recovery Act and the Comprehensive Environmental Response, Compensation, and Liability Act has demonstrated a strong correlation between public acceptance of waste cleanup actions and a lowering of public concerns with their degree of involvement in the decision-input process. Examples of successful actions for RRC—KI officials to explore include (a) open roundtable discussions with stakeholders where information is provided and public concerns are taken seriously; (b) publication of easily understood brochures and Web-based information sources that summarize site history, monitoring results, and cleanup actions; (c) independent public or environmental group monitoring of the environment (e.g., water, air, radiation levels) outside RRC—KI boundaries; (d) site tours for the public and news media; and (e) formation of an advisory committee representing public, business, city government, and other interests.

Managing Institutional Memory. Institutional memory about site history, waste inventories, and remediation efforts is easily lost as contaminated sites are remediated and workers retire or attain other jobs. This could be particularly true as pressure mounts to use the institute's land for urban development purposes. It is recommended that RRC—KI create a permanent, comprehensive, and archival data management and record-keeping system in an accessible form and format to ensure that future site operators or owners understand, for example, which cleanup actions were carried out, why those actions were selected, monitoring results, health and safety records, and which contaminants remain onsite. Otherwise, future generations will struggle to reconstruct today's cleanup decisions and to understand the potential environmental risks left behind.

The loss of institutional memory can be rapid. For example, in the United States, within 2 years after the chemical waste site of Love Canal in New York State was sold, houses and a school were built atop the site, although the transfer deed specifically identified potential health hazards. Years later, a public emergency was declared because of illnesses, odors, and contamination seeping from the ground.

REFERENCES

Novikov, V., ed. 2007. The Nuclear Legacy in Urbanized Areas: Generic Problems and the Moscow Case Study (RR-07-001). Laxenburg, Austria: International Institute for Applied Systems Analysis.

Novikov, V. M., V. V. Lagutov, T. G. Sazykina, Yu. E. Gorlinskii, O. A. Nikolsky, and V. I. Pavlenko. 2005. Assessment of the effect of temporary storage sites for radioactive wastes on the territory of the Russian Science Center Kurchatov Institute on the population and the environment. Atomic Energy 99(2):588-595.

Rastorguev, A., K. Buharin, V. Volkov, D. Tsurikov, Yu. Zverkov, I. Rastorguev, and E. Volkova. 2005. Prognosis of radionuclide contamination spreading on the site of temporary waste storage of RRC Kurchatov Institute. In Proceedings of the International Congress ECORAD'2004: The Scientific Basis for Environment Protection Against Radioactivity, September 8-10, 2004, France, Vol. 40, Supplement 1, pp. 367-370.

Volkov V. G., N. N. Ponomarev-Stepnoi, E. S. Melkov, E. P. Ryazantsev, V. S. Dikarev, G. G. Gorodetsky, Yu. A. Zverkov, V. V. Kuznetsov, and T. I. Kuznetsova. 2003. Status of activities on rehabilitation of radioactively contaminated facilities and the site of Russian Research Center Kurchatov Institute. In Proceedings of the Waste Management 2003 Symposium, Tucson, Arizona, February 23-27, 2003. Tucson: WM Symposia, Inc.

Volkov, V. G., Yu. A. Zverkov, S. G. Semenov, A. V. Chesnokov, and A. D. Shisha. 2009. Remediation of Contaminated Facilities at the Kurchatov Institute. Pp. 99-109 in Cleaning Up Sites Contaminated with Radioactive Materials: International Workshop Proceedings. Washington, D.C.: The National Academies Press.

17

Industrial Nuclear Explosion Sites in the Russian Federation: Recovery and Institutional Monitoring Problems

V. V. Kasatkin, Ye. N. Kamnev, and V. A. Ilyichev, Russian Federal Atomic Energy Agency (Rosatom), All-Russian Research, Design, and Surveying Institute of Production Technology

CLASSIFICATION OF INDUSTRIAL NUCLEAR EXPLOSION SITES CONTAMINATED WITH RADIONUCLIDES

This paper presents a classification of industrial nuclear explosion sites contaminated with radionuclides and covers problems associated with site decontamination and radiation monitoring. There were 81 underground nuclear explosions in the Russian Federation from March 1965 to September 1988. A number of these explosion sites are contaminated with radionuclides as a result of the explosions themselves, the boring of holes leading to the explosion zones, or subsequent technogenic and natural processes. The scope of the problems involved in recovery at such sites and their monitoring will depend on contamination type and current site status.

In terms of radionuclide contamination genesis, nuclear explosion sites may be classified as follows:

1. Sites where there was a planned release of technogenic radionuclides (including alpha-emitters) to the earth's surface during discharge-oriented explosions (the Taiga site) and bucklings (the Kristall site), resulting in the generation of radioactive buildup and a fallout trace.
2. Sites where explosions took an unplanned turn, resulting in the release

of radionuclides that formed a fallout trace, such as occurred following the two botched explosions Kraton-3 and Globus-1. The radiation consequences of these explosions were determined by the channels by which the blast products reached the surface, the time it took for this to occur, and the gas content in the rock at the point where the nuclear device was placed. At the Kraton-3 site, explosion products began erupting through the loading-hole shaft 5 seconds after the explosion and lasted for about 10 minutes. During this explosion in the carbonate reservoir rock, in addition to fission fragment radionuclides, alpha-emitters were also discharged, yielding a fallout trace (based on a commitment dose of 5 mSv) reaching about 30 km in length. At the Globus-1 site, radionuclides began to be discharged 17 minutes after the explosion. Because of the high fume characteristics of the rock (limestone), site contamination with fission-fragment radionuclides along the trace did not exceed several hundred meters despite the long duration of gaseous product (carbon dioxide) outflow (more than 10 days). Radionuclides were discharged through the casing space of the loading hole as a gas-water gryphon, which partially cleansed the gas of nonvolatile radionuclides. A pressure-induced gas eruption from the explosion zone also occurred at the Globus-3 site. In the latter case, the gas eruption began about 12 minutes after the explosion and lasted for 7 hours. This time, the gas flow rate and site contamination level were considerably lower than at the Globus-1 site, since the Globus-3 nuclear device had been placed in clay rock with sandstone layers.

3. Sites contaminated as a result of hole boring in the central explosion zone, particularly in connection with loading-hole recovery. Recovery of loading holes in rock salt resulted in the controlled emission of various forms of tritium and inert radioactive gases, which did not lead to any long-term site contamination. The boring of holes in a water-bearing horizon leading to the central explosion zone resulted in contamination of the immediate site grounds with cesium-137, strontium-90, and tritium entrained in drilling fluid and sludge, as well as stratal water extraction from the explosion zone during hydrodynamic survey works. This was the case with the Globus-1, Globus-2, Kama-1, and Kama-2 sites. At the Kama-1 site, the primary cause of site contamination was the unplanned discharge of carbon dioxide and radioactive water from the loading hole during its recovery. The contaminated zone exceeded 30,000 m^2 in area. It should be noted that the cesium-137 and strontium-90 contamination of the Globus-1 site was mainly caused by hole boring in the explosion zone.

4. Sites where technogenic radionuclides were released to the earth's surface and distributed in the soil as a result of violations of technical procedures. The most typical example in this category is the Grifon site. At this site, a product (crude oil) was extracted from the explosion zone along with radioactive water, which after being separated from the oil was used for repressuring. Leakages in the injection wellheads resulted in the contamination of dozens of wellhead zones with cesium-137, strontium-90, and tritium, while accidental pipe ruptures contaminated areas beyond the immediate drilling sites.

5. Sites where technogenic radionuclides were discharged to the earth's surface due to natural processes. This situation is most typically observed at a number of reservoirs formed in rock salt by nuclear explosions at the Vega site, but it may also be found at many other sites. Such reservoirs tend to decrease gradually in volume. This process leads to radioactive brine extrusion, corrosion failure of wellhead valve threaded connections, and the release of brine on the surface around the wellheads. In addition, brine escape and equipment contamination may result when accumulated water is pumped out of such reservoirs or when borehole cavities are repaired or liquidated.

The above classification is somewhat simplified, as radioactive contamination may be caused by various factors; however, it offers a basis for systematic application of site recovery methods.

DEACTIVATION OF CONTAMINATED TERRITORIES

Covering a site with clean soil is the simplest deactivation method. This technology has been implemented at relatively small remote sites (Globus-3 and, to a certain extent, Globus-2).

At the Kama-1, Kama-2, and Grifon sites, a layer of contaminated soil was removed with bulldozers, loaded into trenches or piles, and isolated with a layer of clean soil. However, it is impossible to restore natural background radiation levels simply by removing soil with heavy equipment. Therefore, after the soil was removed, a layer of clean soil was scattered over the site. Site deactivation work has been under way for several years at the Grifon site, with the radioactive soil being buried in a specially designed subsurface container constructed in compliance with existing rules and regulations. At the Kama-2 site, soil and equipment are collected in subsurface containers arranged as trenches in a thick clay layer above groundwater level. At the Kama-1 site, in view of the high groundwater level, more than 3,000 m^3 of radioactive soil has been collected in a pile, the "physical" protection of which is provided by a clean soil layer 5-20 cm thick and a concrete barrier. As yet, no decision has been made on the future fate of this repository.

The Globus-1 site is the most problematic. It includes contaminated soil areas, subsurface repositories of radioactive soil and fine particulate materials in sand, and subsurface soil areas up to 3 m thick that contain radionuclides and are covered with a sand layer 5-100 cm thick. The site is located on a riverbank and may be affected by spring floods. This may flood the wellhead zone, possibly washing surface sand out into the river. A site recovery project that has been developed, including onsite containment of radioactive products, is not being implemented due to lack of funding (the calculated budget exceeds 36 million rubles in 2002 prices).

The Kraton-3 site includes a well-equipped repository for contaminated soil and equipment that has been in existence for more than 27 years. Currently, it is being renovated in accordance with an approved project. The radioactive trace area is a woodland with uneven terrain marked by granite outcrops. Recently, because of a number of technical and financial issues, the idea of decontaminating the trace was dropped. Radiation warning signs are currently being installed along the boundary of the radioactive trace area (0.5 × 2.5 km in area). The same decision was made regarding the Taiga site, where, apart from warning signs around the perimeter, plans call for cutting trails into the forest and installing additional warning signs at various distances up to 1.2 km from the site.

At the Kristall site, the radioactive trace area and piles were covered with a thick layer of rock refuse from the diamond quarry in 1992 and again in 2006. Beyond the bounds of the site, there may be points of soil contamination with gamma-radiation dose levels only slightly exceeding natural background level.

MONITORING ISSUES

Because of the presence of contaminated areas and radioactive soil repositories, procedures must be established for monitoring all of the industrial nuclear explosion sites, including those where there are currently no signs that any radionuclides have been released. At most of the sites, the radiation situation may be characterized as consistent with natural radiation background levels, which can vary widely. However, the possibility of future radionuclide releases cannot be completely ruled out if well-casing column corrosion and cement degradation occur near water pressure horizons in the explosion zone or beneath the explosion cavity.

As yet, there is no officially established uniform system of radiation monitoring. Most of the sites in operation have monitoring systems supported by the radiation safety services of the respective facility operators. As for the other sites, radiation monitoring is being mainly provided by the Nuclear Safety Laboratory of Rosatom's All-Russian Research, Design, and Surveying Institute of Production Technology. Contaminated sites are monitored periodically, while other sites are checked on a case-by-case basis.

The scope of radiation monitoring depends on the site status and as a rule includes the following:

- Visual survey of changes in site surface topography, including the presence of possible soil subsidence over the explosion cavity and newly apparent water sources (springs, brooks, discharges from loading and other nearby holes, and so forth);
- Survey of gamma-radiation dose rate, automatically mapped and linked with site coordinates

- Analysis of gamma-radiation dose rate dynamics
- Flux density measurement of beta- and alpha-particles in locations where the gamma-radiation dose rate is increasing
- In-field gamma-ray spectrometry of contaminated areas
- Sampling of soil, water, vegetation, fungi, and so forth, and their analysis for tritium, strontium-90, and cesium-137 content

Monitoring types and schedules are set on a project-by-project basis.

18

Comments on Presentation on Industrial Nuclear Explosion Sites in the Russian Federation: Recovery and Institutional Monitoring Problems

Don J. Bradley, Coastal Security Institute, Marine Research Operations, Pacific Northwest National Laboratory

The paper that was presented lays out the issues of the Russian "peaceful nuclear explosions" (PNEs) with respect to radioactive contamination at PNE sites. It is a valuable contribution to the recent literature where it highlights several PNEs and remediation actions being taken or planned. According to Russian documents, there were 124 peaceful nuclear explosions, all underground, with 117 of them being conducted outside the nuclear test site boundaries (Semipalatinsk and Novaya Zemlya). This can be broken down by country as 80 in Russia, 39 in Kazakhstan, 2 each in Ukraine and Uzbekistan, and 1 in Turkmenistan, covering a time span of 23 years (Minatom and MOD, 1996).

The Russian program was quite extensive compared to the U.S. effort in this area, commonly referred to as the Plowshare Program, where 27 nuclear explosion tests were conducted over a period of 11 years. Although the program started with great expectations, many test ideas did not go beyond the planning phase. The potential intended uses included the following:

- Widening the Panama Canal
- Cutting paths through mountainous areas for highways
- Connecting inland water systems
- Creating underground caverns for water, gas, and petroleum
- Generating steam for electricity

- Production of useful isotopes and their recovery
- Connecting underground aquifers
- Stimulating flow of natural gas in "tight" underground formations

In the end, the most promising use proved to be stimulation of natural gas production, which was also an active part of the Russian PNE program. Public opposition to the tests, concerns about contamination, the potential for radioactive gas flaring operations and other environmental hazards, tritium-contaminated gas, and poor economics related to gas production conspired to bring an end to the U.S. PNE program in 1975 (DOE/NV–209-REV 15).

The former Soviet Union's PNE testing program had stronger support, and paralleled the themes of intended uses as seen in the United States, specifically,

- Experiments to develop craters and move earth (such as the Pechora-Kama rivers canal and the village of Udachny in Yakutia-Sakha, and the Chagan River valley in Semipalatinsk Oblast)
- Creation of cavities in salt mines for petroleum and possibly liquid radioactive waste storage
- Enhancement of gas release and control of crude oil flow (such as Ust-Balyk in Tyumen Oblast, Grachevsky deposit in Bashkortostan)
- Underground blasts for seismic probing of the earth's crust and mantle
- Restoration after accidents and fires in gushing oil wells

A map of the locations of the former Soviet Union's PNE program is shown in Figure 18-1.

As a brief summary, the products of a nuclear explosion are distributed in the following places, constituting the initial source term for potential migration of contaminants (National Research Council, 2000):

- The solidified puddle of melted rock in the bottom of the cavity
- On surfaces within the cavity and, in some cases, the borehole
- On surfaces in natural and explosion-induced fractures
- In the atmosphere

While the emphasis seems to be placed on radionuclides, it should be noted that the presence of a chemical source term (such as lead or arsenic) derived from the weapons package, rigging, cables, and instrumentation needs to be addressed as well.

The leaching of radionuclides from the rubble left from the nuclear explosion is an important pathway for tests that were conducted under the water table or in or under perched aquifers. With time, groundwater gradually flows back into the cavity and chimney and comes into direct contact with the radionuclides.

FIGURE 18-1 Map of peaceful nuclear explosions in the former Soviet Union (FSU).
SOURCE: Bulatov, 1993.

Once dissolved, the radionuclides are available for migration through groundwater flow. Mathematical models exist that could be used for the prediction of transport of radionuclides in groundwater. One-dimensional models are probably sufficient, as many values for needed parameters are either lacking altogether or are known imperfectly (National Research Council, 2000).

The approach to site remediation should first start with a definition of what is meant by site cleanup. For underground nuclear test areas, other important steps are as follows (National Research Council, 2000):

- A regional groundwater flow model should be developed, which will provide the initial basis for determining the magnitude of health risks.
- Boundaries will be defined to establish areas of contaminated groundwater.
- Contaminant migration will be estimated and contaminant boundaries will be defined.
- Subsurface contaminants in and around the cavities created by underground nuclear tests will be closed in place, since cost-effective groundwater technologies have not yet demonstrated an ability to effectively remove or stabilize radioactive contaminants.
- Data will be declassified, which will allow a broader range of stakeholders to view the relevant information and increase the likelihood that consensus can be reached on remedial strategies and objectives.

As noted by Kasatkin et al. (2008), the primary issues related to the use of PNEs in Russia fall into the category of surface and near-surface contamination that can spread to surface waters and potentially groundwater systems. Although the level of information presented is very general in nature, it appears that actions have been taken to

- identify selected sites of concern;
- determine the contamination profile for some of the sites;
- initiate and/or complete remediation actions on a few of the sites; and
- where funding was not available or complete, determine the actions that are needed.

It appears that thought has been devoted to the implementation of monitoring systems for PNE sites in Russia. However, it is also noted that there is no officially established system of radiation monitoring—some sites are monitored periodically and others on a case-by-case basis. Certainly this effort needs to be more fully developed.

Illustration of remediation actions are given in the paper by the authors, but only for a few selected sites:

- The Globus-3 and Globus-2 sites, where contamination was covered with clean soil
- The Kama-1, Kama-2, and Grifon sites, where contaminated spoil was removed and placed in trenches or subsurface burial areas, followed by covering with clean soil
- The Kristall site, mostly covered by heavy layers of rocks
- The Globus-1 site, which may be of most concern, as it is located on a river, yet remediation has not been implemented because of lack of funding
- The Kraton-3 site, where remediation appears to have stopped because of technical and financial issues
- The Taiga site, where, other than warning signs, no further remediation is planned

Suggested areas for further elaboration or additions to the report are as follows:

1. It is clear that a complete listing of the sites with associated contamination is needed!
2. Some of the geologic and hydrologic concepts have been translated, but additional clarification would help further understanding by non-Russian audiences. Some examples are

- "discharge-oriented" explosions,
- "bucklings" (see Kristall site),
- "radioactive bulk," and
- "emergency eruption of radionuclides" (Kraton-3 and Globus-1 sites).

3. Although contamination data were shown on selected PNEs in the presentation slides, they were in terms of dose rate. Adding data on isotopic quantities in terms of curies would provide a needed perspective as well as a pathway for risk analysis.
4. While it appears that the authors have looked at remediation options, the issues of surface contamination at PNE sites closely parallels remediation work done both in the United States and in Russia on uranium mill tailings sites. It is suggested that the authors look at the results from these programs and incorporate them as appropriate.

Finally, there appear to be two major concerns related to the completion of remediation work at PNE sites. First, the authors note the lack of funding, which has caused remediation at several sites to either stop or not begin. Second, due to many organizational and management changes in the Russian Federation since the time the tests were conducted, "ownership" issues of the various PNE sites may be hindering remediation progress.

REFERENCES

Bulatov, V. I. 1993. USSR's 200 Nuclear Test Sites: Geography of Radiation Catastrophes and Contamination. Distributed at the Fourth Ex-USSR Nuclear Society Meeting, Nizhniy Novgorod, Russia, June 28-July 2, 1993.

Kasatkin, V. V., Ye. N. Kamnev, and V. A. Ilyichev. 2009. Industrial Nuclear Explosion Sites in the Russian Federation: Recovery and Institutional Monitoring Problems. Pp. 116-120 in Cleaning Up Sites Contaminated with Radioactive Materials: International Workshop Proceedings. Washington, D.C.: The National Academies Press.

Minatom and MOD (Ministry of the Russian Federation for Atomic Energy and the Ministry of Defense of the Russian Federation). 1996. USSR Nuclear Weapons Tests and Peaceful Nuclear Explosions: 1949 through 1990. Minneapolis: Eastview Publications.

National Research Council, Committee on the Remediation of Buried and Tank Wastes. 2000. Long-Term Institutional Management of U.S. Department of Energy Legacy Waste Sites. Washington, D.C.: National Academy Press.

Plowshare Program Web site: *www.osti.gov/opennet/reports/plowshar.pdf.*

United States Nuclear Tests, July 1945 through September 1992 (DOE/NV—209-REV15). Available online at *www.nv.doe.gov/library/publications/historical/DOENV_209_REV15.pdf.*

19

The Past, Present, and Future of the Facilities at Andreev Bay

A. P. Vasiliev, International Center for Environmental Safety, Russian Federal Atomic Energy Agency (Rosatom)

INTRODUCTION

The Andreev Bay facilities were built in the late 1950s and early 1960s for the purposes of (1) nuclear submarine reactor refueling, (2) interim storage of spent nuclear fuel before its shipment to Mayak, (3) storage of solid radioactive waste from the Northern Fleet, and (4) treatment of liquid radioactive waste generated during facility operations. Thus, great quantities of spent fuel and radioactive waste were accumulated on the site during the more than 40 years of its operations. The spent fuel and radioactive waste storage facilities were never repaired, and today they are in very bad or even dangerous condition, presenting a serious radiation hazard to personnel and the environment. Some structures have become unsound, and radioactive substances escape with groundwater to the surrounding land and into Andreev Bay (see Figure 19-1).[1] In May 1998 the government of Russia issued a decree instructing the Ministry of Atomic Energy (formerly Minatom, now the Russian Federal Atomic Energy Agency [Rosatom])

[1]Vasiliev, A. P., V. A. Mazokin, M. E. Netecha, Yu. V. Orlov, and V. A. Shishkin. 2001. Radiation inheritance of Russian nuclear fleet and ecological safety problems relating to utilization of nuclear submarines and rehabilitation of other facilities in the Navy. Pp. 43-46 in Institution of Mechanical Engineers Conference Transactions, Radioactive Waste Management 2000: Challenges, Solutions, and Opportunities. London: Professional Engineering Publishing.

FIGURE 19-1 Main potentially hazardous nuclear and radiation facilities at Andreev Bay.
NOTE: 1—liquid waste processing building (structure 1); 2—spent fuel dry storage units (vessels 2A, 2B, 3A); 3—former spent fuel storage facility (building 5); 4—radioactive waste storage facility (building 6); 5—solid radioactive waste storage facility (7, 7A, 7B, 7B1, 7G, 7D, 67, 67A), open sites (7V, 7E, collector pipes); 6—stationary dock (structure 32).

to resolve the problems associated with nuclear submarine decommissioning and remediation of the former coastal bases, which came to be referred to as temporary spent fuel and radioactive waste storage sites.

PRESENT CONDITION OF THE INFRASTRUCTURE, SPENT FUEL, AND RADIOACTIVE WASTES

Minatom inherited a troublesome legacy. Practically the entire infrastructure at the facilities was destroyed, including the heat, water, power supply, and sewage systems. Most facilities intended for spent fuel and radioactive waste storage had suffered operational damage and were not being used for their designated purposes. The old and new spent fuel storage facilities were particularly dangerous.

Building 5, with two 300 m^3 spent fuel storage pools, began operating in 1962. In 1973 an annex with two 600 m^2 pools was added (see Figure 19-2).

In February 1982 a water level decrease was noticed in one of the pools, which suggested leakage. The same problem appeared later in the second pool. Ice formed by water leaking from the pools appeared on the outer surface of the building wall. It permeated the wall and the foundation, and the outdoor background radiation level reached about 10 mSv per hour. To prevent personnel exposure, the wall at the leakage site was banked on the outside with a deep layer of soil. Radioactive water from the pools seeped into the ground under the building and reached an underground stream that outcropped about 20 m downhill from the building and then flowed into the bay. Later on, work was carried out under a contract with Norway to divert the stream from the building area, but most of the land near its former channel remained heavily contaminated. Attempts to locate and stop the leaks failed, and a decision was made to take urgent steps for providing temporary storage for all fuel kept in Building 5.

Plans called for using three 1,000 m^3 reinforced concrete tanks embedded to a depth of 5 m. Metal tubes 250, 275, and 300 mm in diameter were installed in the tanks, and the space between them was filled with concrete. The tube spacing was designed so the system would remain strongly subcritical even if an accident were to flood the tubes.

Fuel retrieval from the damaged storage facility of Building 5 was by no means an easy job, but ingenious and dedicated naval personnel succeeded in moving all the canisters from Building 5 to the dry storage units (DSUs), with

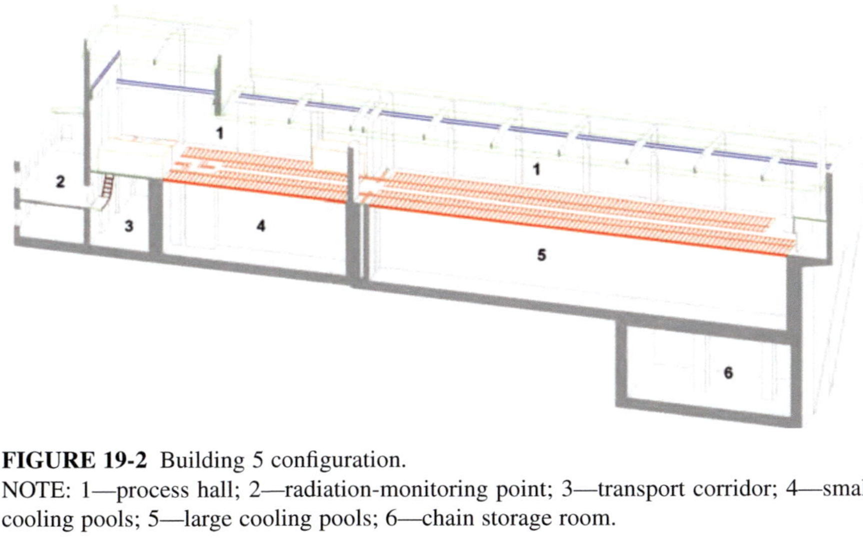

FIGURE 19-2 Building 5 configuration.
NOTE: 1—process hall; 2—radiation-monitoring point; 3—transport corridor; 4—small cooling pools; 5—large cooling pools; 6—chain storage room.

some fuel being taken to Mayak. Arranging spent fuel shipments from the site was significantly slower than unloading the submarines, and all three units were quickly filled. The DSUs and the old-design casks accommodated the fuel unloaded from about 100 submarine and icebreaker reactors.

Three open pads and seven storage facilities are in use for storing solid radioactive waste. The facilities have been in operation for more than 35 years and are no longer leak-tight. The exact quantities, composition, condition, and characteristics of the solid waste, especially those stored in subsurface facilities, are unknown.

Liquid radioactive waste is kept in reinforced concrete tanks lined with stainless steel. Their service time expired long ago; three tanks are no longer leak-tight and are exposed to groundwater penetration. Besides these tanks, liquid waste is also found in the cells of DSUs, as well as in the subsurface solid radioactive waste storage facilities.

Since 2001, substantial work has been carried out to improve the radiation and environmental conditions. Large equipment and transport vehicles (handling equipment, service vessels, and trucks) were dismantled and removed from the land and water areas. The contaminated fragments were placed in temporary packaging and then stored.

To improve working conditions, top priority infrastructure components were provided, such as a changing room, mobile checking and cleaning stations, a decontamination pad, a laboratory, an administrative building with amenities, and an upgraded main water conduit.

Examination of the interim storage site began when it was handed over to Minatom. Information about the main hazards, their locations, activity levels, and conditions is essential for assessing the status of the site and the surrounding environment. The results of these studies served as input data for justification of the investments and decisions on the design details.

Work at Andreev Bay is proceeding in three major directions, with the funds provided by Norway, Sweden, and the United Kingdom, respectively. Some investigations in all the target areas are financed by Rosatom.[2] Norway is financing the ground surveys, including studies on the radioactivity distribution in soil profiles and in groundwater, as well as supporting the infrastructure restoration.[3]

Based on the results of studies carried out by Russian specialists from the N. A. Dollezhal Research and Development Institute of Power Engineering (NIKIET) from 2002 to 2005, radiation-contaminated areas were mapped, territorial zoning was performed to establish controlled access areas, and decontamination

[2]Akhunov, V. D. 2001. High priority projects: Current status of organization and implementation. Presentation made at the Thirteenth Meeting of the International Atomic Energy Agency (IAEA) Contact Experts Group, Oskarshamn, Sweden, November 6-8, 2001.

[3]Panteleev, V. N. 2006. Work performed in 2002-2005 under contracts with the Kingdom of Norway. Voprosy Utilizatsii APL [Issues in Decommissioning Nuclear-Powered Submarines] 2(10):52-57.

plans were made and partly implemented in some hot spots (near the new pier). The work yet to be done involves a comprehensive survey of about 60 percent of the territory, including in-depth exploration of the soil in the area downstream from Building 5, a potential contamination source of the seawater and bottom sediments.

Measurements of specific soil activity are tracked by cesium-137 baseline data (with the measured maximum reaching 9 × 106 Bq/kg) and strontium-90 data (4 × 106 Bq/kg). Gamma dose rates near the points with such activity reach 450 μSv per hour and are as high as 1,000 μSv per hour at the old pier.

Surveys of the liquid radioactive waste storage facilities and open solid radioactive waste storage sites, as well as a feasibility study on radioactive waste management, were carried out under a contract with Sweden.[4] The solid waste storage facilities, where highly radioactive materials are expected to be found and groundwater presence is a strong probability, have not yet been explored.

An inventory of the solid radioactive waste stored in open sites drastically contrasted with previous information about its quantity, with the new data indicating the presence of several times greater amounts of such waste. In addition, gamma background measurements taken at different points on the surface of casks showed large variations (a factor of several tens). This points to considerable nonuniform activity distribution inside the casks and suggests the presence of individual radiation sources. This discovery calls for additional precautions to reduce personnel exposure risk during unloading of the casks.

Radiation distribution on the solid waste storage pads has a highly irregular pattern. For instance, the gamma background at Buildings 7 and 7A is 1,000-3,000 μSv per hour, with the beta contamination of the soil coming to 4,000-6,000 particles per square centimeter per minute. The gamma background here is determined not by the soil contamination but by the radiation from one or possibly two casks placed on the roof of Building 7A. In most other places where solid waste is stored openly, the governing factor is soil contamination. As suggested by experience, the gamma background level may rise sharply after removal of some casks or concrete plates from the ground surface, with loss of the shielding they provided. This should be taken into account in planning operations. The main radionuclides are cesium-137 and strontium-90, but in some places the cobalt-60 contribution may approach 10 percent. Practically no alpha emitters have been discovered.

The spent nuclear fuel storage facilities (DSUs and Building 5) are being surveyed under contracts funded by the United Kingdom.[5] The first exploration

[4]Izmailov, D. M., A. P. Vasiliev, and S. M. Kapinos. 2006. Radioactive waste handling on the Andreev Bay site. Voprosy Utilizatsii APL 2(10):80-85.

[5]Field, D., and V. Aden. 2003. Progress in the SNF management project. Presentation made at the Sixteenth Meeting of the IAEA Contact Experts Group, The Hague, The Netherlands, April 23-25, 2003.

of Building 5 was not undertaken until more than 12 years after it closed. The inspection covered the process hall, the pools, and the chain storage room. Based on the information obtained, a program for comprehensive engineering and radiological survey (CERS) was prepared and carried out in 2005. The main tasks associated with the CERS of Building 5 include the following:

- High gamma background in the chain room (0.3-1.0 μSv per hour) and possible collapse of the piled-up chains dropped from the process hall (pile height approximately 16 m; base area approximately 10 m^2)
- Very high gamma background in the process hall: up to 9.5 mSv per hour above the cantilever beams of the pools, approximately 100 mSv per hour on the bottom of the pools, and up to 600 mSv per hour at individual points; specific activity of samples taken in the hall as high as 1.2×10^9 Bq/kg (for cesium-137); a thick layer of dust on the passage floor and concrete plates
- Cluttered space in the passages and above the pools, which adds to the risk of the piled-up concrete plates and beams collapsing into the deep (6.5 m) pools

Still more dangerous are the conditions in the DSUs. Some cells are filled with water, with activity varying widely from 10^{-7} to 10^{-2} Ci/L. In 2004, NIKIET specialists took the first samples of water at different depths. The activity increases with depth several times for cesium-137 and tens or hundreds of times for strontium-90. Alpha emitters are definitely present. For instance, at the bottom of one cell, the activity is 8.9×10^7 Bq/L for cesium-137, 7.2×10^8 Bq/L for strontium-90, and 5.4×10^4 for alpha emitters. This suggests that fuel is exposed to water and that the latter is picking up not only fission products but also actinides. The spent fuel degradation is promoted by the high content of salts (up to 1,500 mg/L), including chlorides (up to 400 mg/L).

Studies of corrosion processes conducted at Russian research institutions have pointed to the possibility of spent fuel degradation and disintegration into fragments of approximately 200 μm, which may stay inside the cask, forming a uranium-water mixture with a small proportion of the latter. In this case, the distribution coefficient (K_{eff}) for one cask is much less than 1.0.

To rule out the possibility of a chain reaction, the following procedure for spent fuel retrieval from DSU cells was adopted:

- Drainage of the cells by pumping
- Reloading of spent fuel assemblies into new canisters with the help of a standard transfer cask or using a retrieval machine that is being developed

The loaded canisters will be put into TK-18 shipping casks, placed on a buffer pad, and then carried by a special vessel either to Murmansk or to Severodvinsk for subsequent shipment to Mayak.

Construction of an oil terminal in the Kola Gulf near Murmansk will increase the traffic of large oil tankers. It appears advisable, therefore, that spent fuel should be taken from Andreev Bay to Severodvinsk, where a spent fuel transfer point has been set up with U.S. assistance. However, in considering this option, we should remember that spent fuel shipment from Severodvinsk is problematic today, as the bridge across the mouth of the Nikolskoye River does not meet regulatory requirements. A new railway bridge must be built, and the need for such a bridge was justified in the Strategic Master Plan.[6]

At Andreev Bay, spent nuclear fuel is found not only in DSU cells but also in old-design casks, which were kept for a long time on an open pad. Today the condition of the fuel inside the casks is unknown.

REMEDIATION OF THE TEMPORARY STORAGE SITE AND REDUCTION OF RADIATION RISKS

The remediation of the temporary storage site and reduction of the radiation risks are already in progress. The following activities are supported by funds provided by the United Kingdom and Norway.

- The road to the interim storage site has been repaired, the administration building with amenities has been constructed, and the infrastructure is being restored.
- The physical protection system is being set up.
- Plans have been made to install the radiation-monitoring system.
- Two mobile checking and cleaning stations and the required dosimetry equipment have been delivered and placed in service, and a radiochemical laboratory has been set up.
- Surveys of the grounds, buildings, and structures are being carried out.
- A cost-benefit analysis has been prepared for building the spent fuel and radioactive waste handling infrastructure.
- The technical and safety aspects of spent fuel handling options are being studied as a basis for final selection.

The main problem is moving the spent fuel from Andreev Bay to Mayak, which is essential for nuclear safety reasons. For this to be achieved, the following should be developed and implemented for spent fuel handling:

- A handling and transportation process, including the equipment required for unloading spent fuel canisters from DSU cells and casks of types 6 and 11

[6]Sarkisov, A., ed. 2004. Strategic Approaches in Solving Decommissioning Problems of Retired Russian Nuclear Fleet in the Northwest Region. Moscow: Russian Academy of Sciences Nuclear Safety Institute.

- Technology for handling damaged spent fuel, which is currently estimated at no less that 20 percent of the total, with this percentage expected to increase as the fuel remains in cells filled with water of high salt content
- Safe working conditions at all stages of the operations

At all stages, spent fuel handling gives rise to liquid and solid radioactive waste, including high-level waste. The latter will have to be treated in addition to the already existing solid and liquid waste. This means that modern facilities for handling not only spent nuclear fuel but also radioactive waste should be provided on the Andreev Bay site. It is also necessary to provide temporary storage facilities, including a buffer pad for containerized spent fuel storage before it may be loaded on a vessel, as there are no alternative ways to remove the fuel from Andreev Bay, and a place where solid radioactive waste may be stored before it is shipped away to be buried or disposed of in a regional radioactive waste repository. However, neither the site nor the type of regional repository has yet been chosen. This is still another problem that should be resolved soon.

THE FUTURE OF ANDREEV BAY

The future of Andreev Bay is closely related to its past, although the site will not be used for its previous purpose. Nor does it have any apparent prospects for other uses. In the next 10 to 15 years at least, the activities on the site will include spent fuel conditioning and shipment, radioactive waste conditioning, demolition of contaminated buildings and structures, and remediation of the land and water areas. Arrangement of these activities is shown graphically in Figure 19-3.

The cost and timescale of the remediation activities and the waste quantities to be shipped away for storage or disposal depend on the criteria underlying the end-state requirements that will be set for the grounds and structures at Andreev Bay. Considering that people are not expected to live in the area or even stay there for extended periods, "green field" status is not an objective. "Brown field" status is quite appropriate under the circumstances (see Figure 19-3). It is suitable for the arrangement of temporary solid waste storage before this waste is sent to the regional repository. Of course, this option should be justified in terms of environmental safety and economic efficiency following current regulatory requirements and should have the consent of the public and the authorities of neighboring towns and the entire region, as well as of the donor countries involved in activities at Andreev Bay.

Relevant criteria and supplements to current regulations should also be developed and approved in the next few years because of their great influence on the choice of technologies and the scope of work. One example may be given as an illustration: Some European countries, such as France and Sweden, have introduced the category of very low-level waste (VLLW) (0.3 to 100 Bq/g). A simplified cost-saving disposal procedure has been developed for such waste,

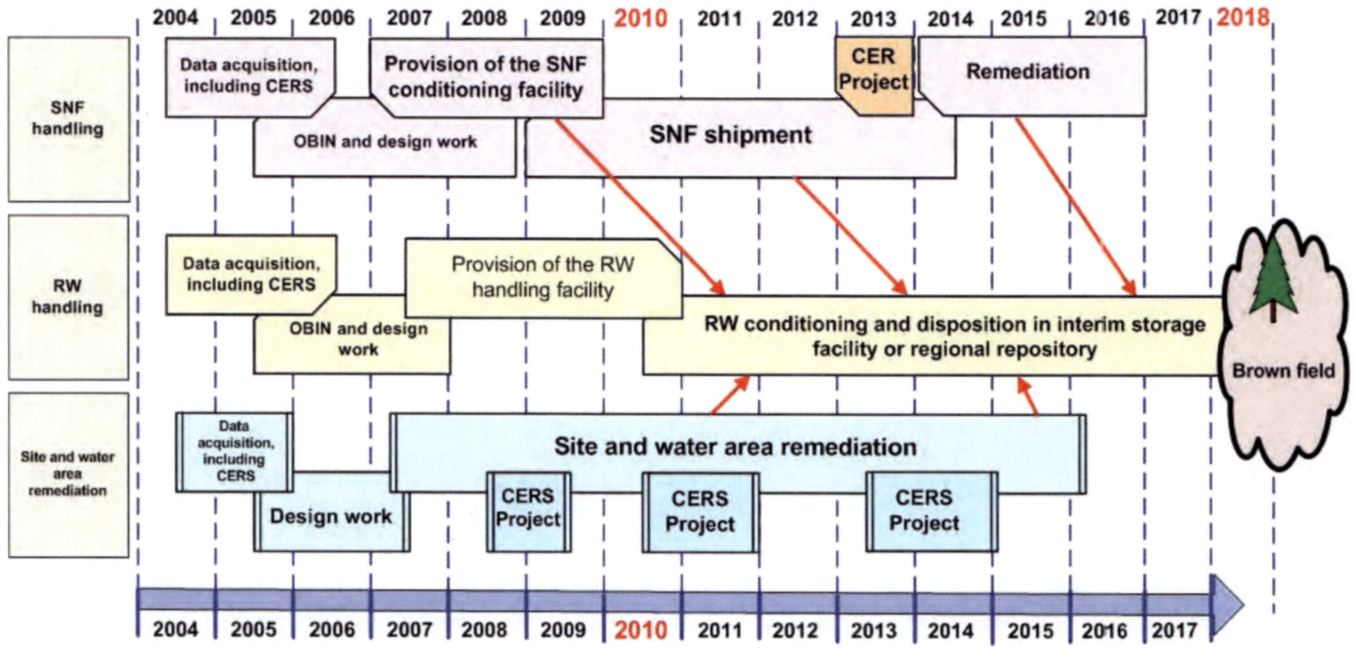

FIGURE 19-3 Work stages in remediation of the Andreev Bay site.
NOTE: SNF—spent nuclear fuel; RW—radioactive waste; CERS—comprehensive engineering and radiological survey; OBIN—cost-benefit analysis.

with the associated safety concerns relating to the impacts on humans and the environment. The bulk of the radioactive waste found on the Andreev Bay site—primarily structural components and soil—belongs to this category. These materials need not be shipped away or containerized. It is simpler and cheaper to put them into surface storage facilities at the same site in keeping with rules and regulations that should be developed, validated, approved by the regulatory authorities, and accepted by the public and nongovernmental environmental organizations.

A proposal for introducing in Russia a waste category similar to the European VLLW was developed within the framework of Strategic Study No. 8 (in accordance with the Strategic Master Plan) in cooperation with representatives of the regulatory authorities.

The condition of Andreev Bay today is a legacy of the Cold War. Remediation of the site may be successfully accomplished through joint efforts by Russia and European countries, from a mutual willingness and ability to live in peace.

20

Environmental Remediation of Spent Nuclear Fuel and Radioactive Waste Temporary Storage Facilities in Gremikha Village: Challenges and Proposed Solutions

Yu. Ye. Gorlinsky, A. Yu. Kazennov, O. A. Nikolsky, V. A. Pavlov, B. S. Stepennov, and A. F. Usaty, Russian Research Center—Kurchatov Institute

CURRENT STATUS OF OPERATIONS IN GREMIKHA VILLAGE

The former Russian Navy technical base in Gremikha village, now called the spent nuclear fuel and radioactive waste temporary storage facility, is situated on the northern coast of the Kola Peninsula. It was built in the mid-1960s to provide technical services for nuclear submarines, with the primary objectives of its work being as follows:

- Refueling pressurized water reactors, as well as storing and removing irradiated fuel assemblies
- Refueling liquid-metal coolant reactors, as well as storing spent removable cores from these reactors
- Receiving, storing, and removing liquid and solid radioactive wastes for disposal or processing

The site covers 150,000 m^2 and has approximately 600 m of shoreline. The terrain at the site is rugged, with the elevation ranging from 0 to 25 m above sea level. In the immediate area, there are no roads, rail lines, or air connections with large cities or even with small settlements. Life support for the site and the closed administrative town of Ostrovnoi is provided by the ship *Klavdia Yelanskaya* and by helicopters. The distance to Gremikha from either Murmansk or Arkhangelsk is approximately 400 km by sea. The lack of a developed transportation network could have negative consequences for future cargo delivery operations.

In 1998 the Russian government terminated operations at the site, and in 2001 it was transferred to Minatom (the Russian Ministry of Atomic Energy, now the Federal Atomic Energy Agency, or Rosatom) for environmental remediation. The objectives are to remove the spent nuclear fuel and the radioactive waste, decommission the facility, and clean up the adjacent territory within a sanitary protective zone until residual contamination reaches acceptable levels. This work is to be carried out with consideration for possible options for further use of the site, provided that safe conditions can be ensured for workers, the public, and the environment.

Circumstances hampering environmental remediation at the site include the following:

- There are no national standards or requirements for environmental remediation of nuclear- and radiation-hazard facilities, nor are there criteria for demonstrating that cleanup goals have been reached.
- There are no approved end-state solutions for the site or decisions on the time allocated to reach that end state, taking into account proposals for further use of its various facilities.
- Hard and urgent work must be carried out in order to provide adequate conditions for spent nuclear fuel and radioactive waste storage, taking into account both the existing damage to the protective barriers and infrastructure and the support of the population.
- Plans for environmental remediation depend on prospects for further development of the town of Ostrovnoi, which is the main center supporting the people involved in this work.
- The positions taken by parties with an interest in Ostrovnoi's future are diverse with differing objectives. Thus, proposed implementation plans for environmental remediation vary considerably.
- Active work is limited to specific seasons because of the climate conditions and availability of transportation connections.
- Previous environmental remediation experience with such facilities is lacking.
- Solutions must be recommended that take into account realistic work dynamics and unstable operating conditions that entail high risks during implementation of planned approaches.

Urgent environmental remediation efforts have already been carried out, including work related to radiation safety of workers during routine operations at the facility, preparation and implementation of comprehensive engineering radiation studies, and partial restoration of the site infrastructure.

Within the Strategic Master Plan for the total decommissioning and remediation of former naval sites in northwestern Russia, proposals have been prepared for environmental remediation of the site. Efforts are being stepped up with additional funding provided by foreign donors as well as by Rosatom. The foreign donors are France (Atomic Energy Commission, CEA); the European Bank for Reconstruction and Development (EBRD); and the European Commission (EC), under the Technical Aid to the Commonwealth of Independent States Program (TACIS). Each of the parties undertakes certain obligations for financing urgent tasks, which include the following:

- The site study component anticipates a comprehensive engineering and radioactive examination of the buildings and structures, site conditions, and offshore waters.
- Preproject documents are to be developed, justifying the selection of optimum engineering solutions for management of spent fuel from Alpha-class nuclear submarines, management of solid and liquid radioactive waste, and the marine infrastructure.
- The container component must provide containers for safe transportation of spent fuel and solid radioactive waste being stored at the site.
- Top priority projects are to focus on urgent tasks related to building repair, equipment modernization, elimination and localization of radiation-hazard hot spots, and so forth.

Previously, Minatom and Rosatom had accomplished the following work:

- A preliminary site study was carried out in 2003-2004, including the grounds, building interiors, structures, and offshore waters. Based on an examination of the results, top priority remediation tasks have been identified in the Strategic Master Plan.
- The site infrastructure was reconstructed to handle the unloading of reactors from Alpha-class submarines. This infrastructure preparation, a top priority activity, has facilitated unloading since 2000.

The following components of the unloading complex were repaired and modernized:

- Dry Dock DD-10
- Reactor refueling building (1A)
- Spent removable core storage facility (1B)
- Universal reloading equipment
- Hoisting gears
- System for physical control of reactors and spent removable cores

In order to heat Alpha-class submarine reactors, a boiler house was designed and commissioned using modern equipment.

In 2005 and 2006, two Alpha-class submarines were placed in Dry Dock DD-10. Spent removable cores were unloaded from these submarines in accordance with technological regulations. They were placed in buildings 1B and 1A, respectively, for interim storage.

According to contracts between the Kurchatov Institute and CEA, Russian organizations carried out the following tasks in 2005-2007, with the resulting information serving as the basis for preproject and project documentation:

- Generalization of results from the preliminary site study, analysis of relevant information regarding availability and conditions of accumulated spent nuclear fuel and radioactive waste, and assessments of the condition of the engineering infrastructure
- Development of the Gremikha information system and generation of a database that will include the results of previous studies
- Purchase and installation of equipment, devices, and materials to create conditions for safe implementation of current radiation-hazard operations, including two modular decontamination facilities and additional radiological equipment
- Completion of a site study aimed at obtaining comprehensive information on both the radiation situation in the buildings, structures, and offshore waters and the engineering conditions of buildings, structures, and the technical infrastructure

CEA has begun to finance activities on 14 projects proposed by Rosatom. Contracts have been signed for five projects, and technical and financial proposals have been prepared for the remainder.

In July 2005, TACIS signed a contract with the Kurchatov Institute entitled "Preliminary Radiological and Engineering Study and Actions for Radiological Protection for Solid Radioactive Waste at Gremikha." Under this contract, additional protective equipment, instruments, and materials were provided to ensure safe performance of current onsite operations. Work completed includes the following:

- Preparation and commissioning of two additional modular decontamination facilities, manufactured with French financing
- Examination of onsite solid radioactive waste facilities to obtain information on the radiation and technical conditions of spent fuel and solid waste in outdoor storage
- Containment of intense radiation sources, resulting in improvement of the onsite radiation situation by more than 6 times (and in some places, up to 200 times)

The site study results are being used to develop projects relating to the transfer of spent nuclear fuel and radioactive waste into interim repositories (EBRD Project 2) for the purpose of improving their storage conditions. In accordance with contracts between EBRD and Russian organizations, the following projects are being implemented:

- **Project 1:** Creation of safe conditions for spent removable cores from Alpha-class submarines
- **Project 2:** Development of a conceptual project for placement of spent nuclear fuel and solid radioactive waste in interim repositories
- **Project 3:** Development of a conceptual project for improvement of spent nuclear fuel storage conditions for the nuclear submarine in Building 1
- **Project 4:** Improvement of the current system of physical protection at the site
- **Project 9:** System of radiological monitoring and emergency preparedness

ENVIRONMENTAL REMEDIATION STRATEGY AND END STATE

The Gremikha site is a radiation legacy of the former USSR. The strategy of environmental remediation at the site is based on several related objectives:

- Identification of strategies for reaching the facility's end state and specification of positive and negative features
- Determination of the main stages and principles of work
- Identification of stakeholders and the basis of their positions
- Specification of principal factors in justifying selection of the most acceptable facility end-state options
- Selection of the most acceptable end-state strategies for the site based on comprehensive assessments of strategy options
- Time and cost evaluations for strategy implementation

The environmental remediation program at the site must consist of interim stages. The remedial strategy must ensure that the situation at the facility will be controllable, stable, protected, and safe at each stage. Security and safety levels for the site as a nuclear- and radiation-hazard facility must increase as each succeeding stage is reached.

A target goal for environmental remediation of the separate storage site facility is as follows (Bylkin et al., 2007):

- Conversion of the facility to comply with regulatory requirements
- Removal of spent nuclear fuel and radioactive waste from the site and subsequent decommissioning of the facility

- Remediation of the adjacent area within the sanitary-protective zone to the extent that residual contamination levels acceptable to the stakeholders are achieved (Gorlinsky et al., 2008)

The environmental remediation work may be divided into four stages:

- **Stage 1** (2005-2010): Urgent facility conversion efforts to achieve conditions meeting regulatory requirements
- **Stage 2** (2008-2015): Removal of spent nuclear fuel and radioactive waste from the site
- **Stage 3** (2010-2020): Decommissioning of the facility and removal of secondary radioactive waste generated
- **Stage 4** (2018-completion): Decontamination of buildings and structures and remediation of adjacent land and water areas within the sanitary-protective zone

Site end-state options and strategies for achieving them are presented in Table 20-1. They differ in regulatory control levels after the end state is reached. The strictest regulatory requirements apply to spent nuclear fuel and radioactive waste storage facilities and to radioactive waste disposal facilities. "Green field" status means that the site would be exempt from regulatory control.

Given this information, the stages with time lines of no more than 10 to 15 years are considered the most justifiable and possible. Further predictions must include hypothetical options based on general wishes as well as declared or potential intentions of stakeholders.

The factor "compliance with actual legislation, government policy, and international obligations" serves as a restriction in implementation of some strategies. Only if this factor were minimized would such strategies be acceptable, as it is difficult to balance the end state after remediation with modern regulatory requirements.

For "spent nuclear fuel and radioactive waste storage facility and/or disposal facility for secondary radioactive waste," which is the end state for the strategy "conversion and facility disposal in situ," regulatory requirements restrict spent fuel and radioactive waste storage facility possibilities. Compliance with these requirements plays a key role in site selection for the facilities for longer term storage and disposal of spent fuel and radioactive waste. To lower this bar when implementing the strategy, the supervisory authority must make special decisions.

Turning to the end state "brown field: industrial (non-radiation-hazard) facility site," this is the end state for the strategy "facility liquidation and site remediation." Implementation of this strategy is hampered by the modern regulatory base having no radiation safety requirements for workers at industrial enterprises (nonradiation-hazard facilities) situated within contaminated areas. In this case,

TABLE 20-1 Strategies for Environmental Remediation at the Gremikha Site

End State	Requirements for Safety Level	Strategy
Current conditions	Compliance with requirements for nonradiation-hazard facilities and for management and operation of spent fuel and radioactive waste storage and disposal facilities	**No action:** Maintenance of facility in current condition without change
Spent fuel and radioactive waste storage facility and/or disposal facility for secondary radioactive waste		**Conversion and temporary operation of facility:** Facility conversion and use according to current function for spent fuel and radioactive waste management and for interim (technological) storage
		Conversion and facility disposal in situ: Facility conversion, spent fuel and radioactive waste removal, and disposal of remaining materials with fixed contamination (secondary radioactive waste) in situ
Brown field: site for a radiation-hazard facility	Full compliance with requirements for radiation-hazard facilities	**Change of facility function:** Spent fuel and radioactive waste removal and subsequent urgent partial dismantlement of the building, as well as decontamination of the building and the site in order to build a new radiation-hazard facility
		Change of facility function: Same as above, but with delayed partial dismantlement
Brown field: industrial (nonradiation-hazard) facility site	Release from regulatory control due to restrictions on site use	**Facility liquidation and site remediation:** Spent fuel and radioactive waste removal and subsequent urgent full dismantlement of structures and radioactive waste removal, as well as remediation of the adjacent area in order to build a new nonradiation-hazard facility
		Facility liquidation and site remediation: Same as above, but with delayed full dismantlement
Green field	Full release from regulatory control	**Facility liquidation and site renovation:** Spent fuel and radioactive waste removal and subsequent urgent full dismantlement of structures and radioactive waste removal, as well as remediation of adjacent area
		Facility liquidation and site renovation: Same as above, but with delayed full dismantlement

regulation is based on individual decisions and covers particular conditions, for example, operation of industrial plants outside areas contaminated with radiation following the Chernobyl accident.

The green field end state is worthy of notice along with its related strategy "facility liquidation and site renovation." In considering a separate facility, this strategy looks tolerable enough. Here, interference with facilities located at the same industrial site was not taken into account, even though the release from regulatory control of a single site at a radiation-hazard facility is unlikely to be justifiable.

Analysis of environmental remediation options has demonstrated that for the next 10 to 15 years, the most acceptable strategy is to "change the facility function." For the Gremikha site as a whole, this strategy includes urgent removal of spent nuclear fuel and radioactive waste and subsequent modification of the site facilities into new radiation-hazard facilities that would provide technical support for remediation work, which could take not more than 10 to 15 years. Two other strategies compete with this one:

- "Conversion of Gremikha site facilities and disposal in situ"
- "Liquidation of facilities and site remediation," which would allow for building on the site an industrial facility that does not involve radiation hazards

The regulatory and legislative base hampers implementation of these strategies. Nevertheless, if special actions and documents were to expand this legislative base to permit implementation of these strategies, they could be considered acceptable for the site.

TECHNICAL SOLUTIONS FOR ENVIRONMENTAL REMEDIATION

Analysis of the Current Radiation Situation

Technological facilities situated within the sanitary protective zone and the radiation-monitoring area are practically clean. However, about 60 percent of the controlled-access area has soil contaminated with cesium-137. The area under the rock near building 1B, the perimeter of building 1A from the side of the rock, and the area close to the liquid radioactive waste storage facility are contaminated zones. The soil there is considered radioactive waste. The total volume of contaminated soil classified as solid radioactive waste (low-level and some medium-level) is preliminarily estimated as about 500 m^3. In general, the radiation situation at the site permits specialists with appropriate permission to work at all site facilities if they observe radiation safety norms according to Radiation Safety Standards-99.

Seawater contamination discovered during a preliminary study of the coastal strip and in water samples taken offshore near the site is due both to radioactivity entering with rainwater from the site and to many years of reception and storage of radioactive waste at the site. Cesium-137 content in bottom sediments has been measured at approximately 10^3 Bq/kg and in seaweed at about 60 Bq/kg, while cobalt-60 concentration has been up to approximately 0.7×10^3 Bq/kg. Contamination of bottom sediments is localized.

On the whole, analysis of the waters near the site based on direct measurements shows that cesium-137 and cobalt-60 content in the seawater and bottom sediments is at the level of background values adopted for this region. The same results confirm an absence of radiation contamination of offshore waters of the Barents Sea with man-made radionuclides cesium-137 and cobalt-60 as a result of site activities.

In the surveillance area (land extending to a 10-km radius from the site), radiation monitoring of the environment is sufficient to assess whether the main control parameters exceed radiation safety standards or control levels, with the exception of monitoring of radionuclide activity concentrations in the near-earth atmosphere. In order to receive such information, special radiation and hygienic monitoring is to be performed along with sensitive sample measurement methods.

The results of radiation monitoring near the site serve as evidence that it is impossible to find impacts of site activities on the environment and the public within the surveillance area.

Preliminary Assessment of the Environmental Impacts of Site Remediation Scenarios

The following types of possible radiation impacts on the environment under routine use of buildings and structures intended for management of spent nuclear fuel and radioactive waste being accumulated within the site should be noted:

- Radioactive gas and aerosol discharges (fission products, radiolysis products, and so forth)
- Releases of storm-water runoff into the open hydrographic network with permissible contamination levels

Concentrations of cesium-137 and strontium-90 radionuclides in the near-earth air layer within the sanitary protective zone and the surveillance area showed instability, ranging from 37 to 3 Bq/m^3. To obtain more unbiased data, stationary facilities for monitoring radionuclide activity concentration in the atmosphere must be established.

The existing site intended for solid radioactive waste operations will be a key source of uncontrolled gas and aerosol releases. Releases from buildings both under restoration and being newly constructed will be monitored. Source parameters and characteristics of releases will be specified.

Assessments of the situation within the sanitary-protective zone and the surveillance area and planned technological actions directed to wastewater cleaning and gas purification show that implementation of remediation work at the site will not necessarily change the radiation situation considerably. During large-scale remediation, the probability of an accident occurring will increase. However, the level of potential accident-related public exposure will not require any fundamental actions to be taken to protect the public. To ensure valid monitoring of the radiation situation, some stationary facilities should be established to track activity concentrations of radionuclides in ambient air.

AMOUNT OF RADIOACTIVE WASTE BEING GENERATED

Solid Radioactive Waste

The main solid waste stream is located within the temporary solid waste storage facilities, located in Building 19 and on the site between Building 1B and the dock. The waste weight is 450 metric tons and waste volume is 800 m^3.

In addition, remediation efforts are expected to produce approximately 2,500 metric tons (about 1,500 m^3) from floating vessel dismantlement; empty spent fuel containers and covers; waste generated during liquid waste cementation; and waste due to disposal of contaminated equipment and rehabilitation of contaminated soil, concrete structures, and so forth.

A considerable amount of additional solid waste (approximately 500 m^3 or 1,100 metric tons) is expected to be created through rehabilitation of soils and concrete structures. In most of this waste, radionuclide concentrations will meet standards for low-level waste. It would be expedient to apply technology to reduce this volume in order to remove it from the site to a longer term storage facility or to transport it to local industrial waste landfills. There are some approved technologies developed for this purpose, which must be adapted to conditions at the site.

Liquid Radioactive Waste

Approximately 300 m^3 of liquid waste is present in liquid waste storage facilities, closed areas of Building 1, and spent nuclear fuel containers at the onsite temporary solid-waste storage facility. An additional volume of about 300 m^3 is expected to be created as a result of decontamination of facilities and floating vessels.

SOCIOECONOMIC BACKGROUND AND DEVELOPMENT PERSPECTIVES

The Gremikha site is situated in the closed administrative city of Ostrovnoi, which receives special funds allocated for such areas. Most of the approximately

350 million rubles in grants received annually are for public utilities, transport services, and communications with Murmansk. However, the government has already prepared a draft directive on revoking the special status and grants.

About 3,000 citizens live in Ostrovnoi. Periodically they receive support for resettlement to the "mainland." Thus, population is decreasing, as is the number of people employed at organizations performing remediation at the site. Because of the remoteness and lack of surface transport, costs of public utilities are about 10 times higher than the average for Russia as a whole.

MAIN REMEDIATION OPERATIONS

A preliminary analysis of operations showed that more than 50 projects are to be implemented in the course of remediation at the site. These activities include the following:

- Analysis of the site and its facilities based on available documentation
- Institution of safety measures for the site study
- Study of the site and determination of the characteristics of its facilities
- Prevention of significant migration of radionuclides into offshore waters and isolation of powerful sources of gamma radiation to reduce the occupational dose level
- Development of management proposals
- Development of projects for the management of spent fuel, spent removable cores, solid and liquid radioactive waste, and contaminated soil (including development of special infrastructure units: storage facilities, containers, handling equipment, and robotics)
- Preparation of the infrastructure for reconstruction of roads, building of switching and buffer areas, building of storage facilities, purchase of equipment and containers, preparation of agreements on use of containerships, and so forth
- Preparation of spent fuel, spent removable cores, and radioactive waste for removal (examination, sorting, placement in containers and other packaging, loading of containers, and preparation for removal)
- Removal of spent fuel, spent removable cores, solid and liquid radioactive waste, and contaminated soil from the site
- Disassembly of contaminated and special equipment and structures and their dismantlement or removal
- Decontamination of buildings intended for conversion
- Remediation of the site and offshore waters

A preliminary analysis of the technology associated with overall spent nuclear fuel repackaging and its removal in transport containers has been carried out. Such technology (considering the tonnage of possible containerships and

TABLE 20-2 Work Categories for Environmental Remediation at the Gremikha Site

Category of Work	Principal Elements According to Operations	Note
Preliminary work	Global site study; creation of required radiation safety conditions for workers; provision of personal protective equipment, decontamination facilities, measurement instruments, and primary robotics and manipulators	
Pressurized water reactor spent fuel management	Examination of irradiated fuel rods (classification as conditioned, C, or damaged, D), pumping of water from containers, removal of C-rods by the ship *Lotta* and D-rods by containership from the site	
Spent removable core management	Order placement, production, and delivery of special containers for spent removable cores, their removal from the site and transport to the Scientific Research Institute of Atomic Reactors (NIIAR), building of infrastructure at NIIAR for spent core management	
Solid radioactive waste management	Formation and provision of technological conditions for solid waste repackaging (sites, repositories, handling equipment, containers)	
Liquid radioactive waste management	Processing of liquid waste into solid form using the Potok unit (cementation)	
Floating vessel management	Dismantlement of Floating Vessel 167 into solid radioactive waste and Floating Vessel 175 into metal scrap	
Environmental remediation of site	Removal of contaminated soil from five locations, separation of soil to minimize the amount of solid radioactive waste to be removed from the site, remediation of buildings, elimination of temporary solid and liquid waste storage facilities	Approval of remediation norms by Federal Medical-Biological Agency
Environmental remediation of offshore waters	Implementation of site study recommendations after removal of spent fuel and radioactive waste from the site and remediation, if necessary, of bottom sediments	

climate-related restrictions) will make it possible to remove all of the spent fuel in the course of 4 years.

In order to make comparative evaluations of the labor intensiveness of the various operations, they have been classified into eight groups, as shown in Table 20-2.

COSTS OF REMEDIATION SCENARIOS

An expert assessment of the expected costs of completing all remediation projects at the Gremikha site (about 50) is approximately 200 million euros. Twelve are considered as priority projects; they are already being implemented and will be finished by 2010. Total cost data relating to the projects are given in Table 20-3.

TABLE 20-3 Remediation Costs

		Cost		
Project Status	Number	Million Euros	Million Rubles	Years
Priority	12	~33	~1,150	2005-2010
Planned	43	~155	~5,250	2006-2025
Total	55	~187	~6,400	2005-2025

CONCLUSIONS

1. Donor countries are providing active financial and technical support for this remediation activity. CEA (France), EBRD, and EC (TACIS) are the main contributors.

2. The following should be recognized as the most acceptable general strategy for environmental remediation at the site:

- Successive conversion of separate facilities (putting them into compliance with regulatory requirements) and temporary continuation of current facility functions and usage
- Urgent removal of spent nuclear fuel and primary radioactive waste
- Decommissioning of facilities and creation of radiation technological facilities for interim management of secondary radioactive waste

3. The assessment showed that environmental remediation work at the site could take 12 to 20 years, depending on various technologies that could be used for spent fuel and radioactive waste management.

REFERENCES

Bylkin, B. K., Yu. Ye. Gorlinsky, V. A. Kutkov, O. A. Nikolsky, V. I. Pavlenko, Yu. V. Sivintsev, and B. S. Stepennov. 2007. Multifactor analysis application for option selection of end state and environmental remediation strategy for spent nuclear fuel and radioactive waste temporary storage site in Gremikha village. Preprint IAE—6456/3. Moscow: Russian Research Center—Kurchatov Institute (in Russian).

Gorlinsky, Yu. Ye., V. A. Kutkov, and N. K. Shandala. 2009. Criteria for environmental rehabilitation of the temporary storage site for spent nuclear fuel and radioactive waste in Gremikha village. Pp. 152-160 in Cleaning Up Sites Contaminated with Radioactive Materials: International Workshop Proceedings. Washington, D.C.: The National Academies Press.

SUPPORTING BIBLIOGRAPHY

Akhunov, V. D., Ye. O. Zhukov, V. A. Pavlov, B. P. Papkovsky, Ye. N. Samarin, B. S. Stepennov, and N. S. Khlopkin. 2005. The coastal base in Gremikha village: Conditions, problems, and proposals for management of irradiated fuel rods from pressurized water reactors. Bulletin on Issues of Nuclear Submarine Dismantlement 4(8) (in Russian).

Crabol, B. 2006. Site specific levels for remediation: CEA experience. Presentation at the Contact Expert Group Workshop on Strategic Aspects of Radioactive Waste Management and Remediation of Contaminated Sites, April 26-27, 2006, Balsta, Stockholm, Sweden. Available online at *www.iaea.org/OurWork/ST/NE/NEFW/CEG/documents/ws042006_17E.pdf.* Additional papers and workshop agenda available at *www.iaea.org/OurWork/ST/NE/NEFW/CEG/ceg_ws042006.html.*

Federal State Unitary Enterprise InterSEN. 2002. Hygiene requirements for safety and nutritional value of foodstuffs. Sanitary-epidemiological regulations and norms, SanPiN 2.3.2.1078-01. 2002. Moscow: InterSEN, 168 pp. (in Russian).

Federal State Unitary Enterprise for Radioactive Waste Management (SevRAO). 2004. Annual Report of the Industrial and Sanitary Laboratory for 2004 on ZB 1, OB 2, SevRAO. Snezhnogorsk, Murmansk Oblast: CGSEN–120, 26 pp. (in Russian).

Gorlinsky, Yu. Ye., O. A. Nikolsky, B. P. Papkovsky, Yu. V. Sivintsev, B. S. Stepennov, et al. 2005. Gremikha Coastal Maintenance Base: Conditions, problems, and prospects for environmental remediation. Izvestiya Akademii Nauk: Energetika, No. 5 (in Russian).

Gorlinsky, Yu. E., O. A. Nikolsky, Yu. V. Sivintsev, B .S. Stepennov, et al. 2005. Challenges of the environmental remediation of the former Russian Navy coastal maintenance base in Gremikha village in terms of development predictions and tendencies of the closed administrative city of Ostrovnoi, Murmansk Oblast. Bulletin on Issues of Nuclear Submarine Dismantlement 4(8) (in Russian).

International Atomic Energy Agency (IAEA). 2005. Selection of Decommissioning strategies: Issues and Factors. IAEA TECDOC-1478. Vienna: IAEA.

IAEA. 2006. Release of sites from regulatory control on termination of practices. Safety Guide No. WS-G-5.1. Vienna: IAEA.

Kurchatov Institute. 2003. Analysis of conditions, development, and implementation of proposals on urgent actions to ensure environmental remediation of the Gremikha facility. Moscow: Russian Research Center—Kurchatov Institute.

Kurchatov Institute. 2006. Implementation of total engineering and radiation study of marine areas at the spent nuclear fuel and radioactive waste temporary storage site in Gremikha village (SevRAO OB 2 area). Moscow: Russian Research Center—Kurchatov Institute.

Kutkov, V. A. 2007. Evolution of the system of radiation safety guarantees in light of new ICRP and IAEA Recommendations. ANRI 1(48):2-24 (in Russian). Available online at *www.doza.ru/anry/archive/2007_1.shtml.*

Kutkov, V. A., V. V. Tkachenko, and V. P. Romantsov. 2003. Radiation safety of nuclear power station plant personnel. Textbook, V. A. Kutkov, ed. Moscow—Obninsk: Atomtekhenergo and Obninsk State Technical University of Nuclear Energy, 344 pp. (in Russian).

Pillette, L. 2003. Experience in site characterization and cleanup at the CEA Center of Fontenay-aux-Roses. Presentation at the Contact Experts Group Workshop on Remediation of the Gremikha Site, October 29-31, 2003, Cadarache, France. Available online at *www.iaea.org/OurWork/ST/NE/NEFW/CEG/documents/ws102003_pillette-e.pdf.* Additional papers and workshop agenda available at *www.iaea.org/OurWork/ST/NE/NEFW/CEG/ceg_ws102003.html.*

Russian Federal Medical-Biological Agency and Norwegian Radiation Protection Authority. 2005. Report on Research Effort "Radiological Impact Assessment on the Population at Routine and Beyond Design Operation Mode of Temporary Storage Sites for Spent Nuclear Fuel and Radioactive Waste in SevRAO Branches during their Operation, Decommissioning, Remediation, and Delicensing." Task 2, Project 2. Moscow, 50 pp.

Sivintsev, Yu. V., S. M. Vakulovsky, A. P. Vasiliev, et al. 2005. Radioecological Consequences of Radioactive Waste Flows in the Seas around Russia ("White Book—2000"). Moscow: IzdAT (in Russian).

Sivintsev, Yu. V., V. L. Vysotsky, R. I. Kalinin, V. G. Aden, and A. P. Vasiliev. 2006. Quantitative criteria for the site remediation of coastal technical bases. Nuclear Energy 101(1):35-49 (in Russian).

State Committee on Sanitation and Epidemic Control (Goskomsanepidemnadzor). 2005. Radiation hygiene certificate OB No. 2 for SevRAO in 2004. 2005. Ostrovnoi, 4 pp. (in Russian).

21

Criteria for Environmental Rehabilitation of the Temporary Storage Site for Spent Nuclear Fuel and Radioactive Waste in Gremikha Village*

Yu. Ye. Gorlinsky, V. A. Kutkov, and N. K. Shandala,
Russian Research Center—Kurchatov Institute

To ensure radiation safety for personnel and the public during the conversion and environmental rehabilitation of the temporary storage site for spent nuclear fuel and radioactive waste in Gremikha village, a radiation safety strategy must be in place that, if followed, would define the following:

- Radiation protection criteria for the planning of work
- Radiation protection criteria for the conduct of work
- Criteria for assessing radiation safety status during the conduct of planned work and at its completion

The regulatory documents in effect in the Russian Federation consider the following two categories of ionizing radiation sources as subject to radiation safety regulation:[1]

---

*Translated from the Russian by Kelly Robbins.

[1]Kutkov, V. A., B. A. Bezrukov, V. V. Tkachenko, V. P. Romanov, I. V. Dolzhenkov, V. N. Lebedev, and V. I. Petrov. 2002. Fundamental principles and requirements of regulatory documents in the practice of ensuring radiation safety at nuclear power plants: A training handbook. V. A. Kutkov and B. A. Bezrukov, ed. Moscow: Rosenergoatom, 292 pp.

Radiation safety norms (RSN-99): Hygiene regulations SP-2.6.1.758-99. 1999. Moscow: Ministry of Health of the Russian Federation.

Basic sanitary rules for ensuring radiation safety of personnel and the population (BSRERS-99): Sanitary rules SP-2.6.1.799-99. 2000. Moscow: Ministry of Health of the Russian Federation.

1. Technogenic radiation sources specially created for their useful application or sources that are by-products of such activity
2. Natural radiation sources covered by the Radiation Safety Norms and Rules

In using the term *ionizing radiation source* or simply *source*, we follow the common practice in radiation protection and safety by which *source* means a radioactive substance or device emitting or capable of emitting ionizing radiation and covered by the Norms and Rules.[2] Everything that could cause irradiation during emission of ionizing radiation or discharge of radioactive substances or materials is considered to be a source. For example, substances that emit radon are sources that exist in the environment, a gamma-radiation sterilization device is a source used in practical operations to preserve food products, and a nuclear power plant is a source in practical operation to produce electricity using nuclear energy. In the application of the Norms and Rules, complex devices or several devices located in one place or at one site may under appropriate circumstances be considered as a single source. In providing radiation protection for people and ensuring their safety, any source is considered as a source of harm and danger. The harmfulness of a source is determined by the real level of radiation associated with it. The danger is determined by the potential exposure that might result if the source goes out of control and leads to an accidental irradiation capable of having substantial radiation effects.[3]

The foundation for ensuring radiation safety for operations involving technogenic sources is the licensing of such activities, which entails the presence of a license holder, a legal entity bearing full responsibility for operations involving the technogenic source. As a rule, such a source is created for a purpose in the form of a device or unit that makes the most effective use of the ionizing radiation created; therefore, such a source is fundamentally a radiation hazard. The regulatory requirements applied to operations involving a technogenic source are primarily aimed at ensuring that such a specially created source is managed to prevent it from going out of regulated control.[4] In this regard, the results of

[2]International basic safety standards for protection against ionizing radiation and for the safety of radiation sources: Safety series No. 115. 1996. Vienna: International Atomic Energy Agency (Russian edition, 1997).

[3]Kutkov, V. A., B. A. Bezrukov, V. V. Tkachenko, V. P. Romanov, I. V. Dolzhenkov, V. N. Lebedev, and V. I. Petrov. 2002. Fundamental principles and requirements of regulatory documents in the practice of ensuring radiation safety at nuclear power plants: A training handbook. V. A. Kutkov and B. A. Bezrukov, eds. Moscow: Rosenergoatom, 292 pp.

Kutkov, V. A., V. V. Tkachenko, and V. P. Romantsov. 2003. Radiation safety for nuclear power plant personnel: A training handbook. 2003. V. A. Kutkov, ed. Moscow-Obninsk: Atomtekhenergo and Obninsk State Technical University for Nuclear Power Engineering, 344 pp.

[4]Op. cit.

Kutkov, V., O. Kochetkov, and A. Panfilov. 2002. Strategy of control at source as a base for protecting workers against risks arising from exposure to ionizing radiation in the Russian Federation.

dosimetric monitoring of radiation exposure of the public and personnel from the device in question are an indicator of its safe status.[5]

The main characteristic of natural ionizing radiation sources is the degree of harmfulness associated with them. Their effect on humans is of a prolonged nature, extended over time. In view of the low specific activity of radioactive materials that the Norms and Rules place in this source category, even if they should go out of control they are incapable of producing substantial doses from accidental exposure.[6]

In considering radiation sources of these two types, the system for regulation of radiation safety formulates its requirements for operations involving them depending on their characteristic danger-harm ratio. For example, annual public exposure from potentially dangerous technogenic radiation sources is limited to a maximum dose of 1 mSv per year, while the total exposure from natural sources for the same period is deemed acceptable if it does not exceed 5 mSv per year. In addition, such exposure is considered as an addition to the existing natural background radiation, the dose from which, as a rule, ranges from 2.5 to 10 mSv per year.[7]

There is also a third type of source along with the sources noted above: radioactive contamination dispersed in the environment as a result of practical activities that have already terminated. As a rule, the individuals or entities responsible for such contamination are not to be found, while the sources themselves are part of the "radiation legacy." Among other aspects, exposure of the affected population and personnel linked with the radiation legacy includes

- exposure from global radionuclide fallout resulting from nuclear weapons tests, and
- exposure at contaminated sites of decommissioned military facilities that currently fail to meet requirements applied to civilian radiation-hazard facilities.

In Occupational radiation protection: Protecting workers against exposure to ionizing radiation. Proceedings of an international conference, Geneva, August 26-30, 2002. Vienna: International Atomic Energy Agency, pp. 39-44. Available online at *www-pub.iaea.org/MTCD/publications/PDF/Pub1145_web.pdf*. Accessed August 16, 2007.

[5]Op. cit.

[6]Radiation safety norms (RSN-99): Hygiene regulations SP-2.6.1.758-99. 1999. Moscow: Ministry of Health of the Russian Federation.

Basic sanitary rules for ensuring radiation safety of personnel and the population (BSRERS-99): Sanitary rules SP-2.6.1.799-99. 2000. Moscow: Ministry of Health of the Russian Federation.

[7]United Nations Scientific Committee on the Effects of Atomic Radiation. 2000. Sources and effects of ionizing radiation: 2000 report to the General Assembly with scientific annexes. Volume 2: Effects, Annex B: Exposure from natural radiation sources. New York: United Nations. Available online at *www.unscear.org/unscear/en/publications/2000_2.html*. Accessed August 16, 2007.

Exposure from such sources is not regulated, yet the sources themselves represent a burden of which we must rid ourselves. Key problems of ensuring radiation safety while eliminating the radiation legacy of the former USSR currently remain unresolved. This circumstance has a substantial effect on the evaluation of the acceptability of a number of end states for sites and facilities under environmental rehabilitation. The fact that there are radiation sources that do not fall within the scope of the domestic regulatory base and that are unregulated is a problem for ensuring radiation safety while eliminating the radiation legacy, a problem that cannot go unresolved.[8]

Limiting the impact of the radiation legacy on the population is not a new problem.[9] The International Commission on Radiological Protection (ICRP) devoted a special publication to this problem,[10] and the recommendations it contained formed the basis for a safety guide issued by the International Atomic Energy Agency (IAEA) in 2006.[11]

In the near future, plans call for the release of new ICRP recommendations and subsequently new IAEA regulations. It is expected that their introduction at the national level will be complete by 2015.[12] Thus, the time period for the conversion of the Gremikha temporary storage site facilities will coincide with the period during which the new regulatory base for radiation safety will be put into practice. One aspect of the new ICRP recommendations that could have a substantial impact on ensuring radiation safety during the decommissioning of the

---

[8]Sivintsev, Yu. V., V. L. Vysotsky, R. I. Kalinin, V. G. Aden, and A. P. Vasiliev. 2006. Quantitative criteria for rehabilitating the territories of shore technical bases. Atomic Energy, vol. 101, 1st edition, pp. 35-49.

Bylkin, B. K., Yu. Ye. Gorlinsky, V. A. Kutkov, O. A. Nikolsky, V. I. Pavlenko, Yu. V. Sivintsev, and B. S. Stepennov. 2007. Application of multifactor analysis in the selection of end state options and environmental rehabilitation strategies for the temporary storage site for spent nuclear fuel and radioactive waste in Gremikha village. Preprint IAE-6456/3. Moscow: Russian Research Center—Kurchatov Institute.

Shandala, N. K., M. F. Kiselev, M. K. Sneve, et al. 2006. Radiation ecology regulation under conditions of rehabilitation work at SevRAO. Pp. 184-186 in Modern problems of ensuring public radiation safety: Proceedings of a scientific and practical conference, St. Petersburg, December 4-7, 2006. St. Petersburg: Federal Monitoring Service for the Protection of Consumer Rights and Human Welfare.

[9]Moiseev, A. A. 1985. Cesium-137, the environment, and man. Moscow: Energoatomizdat.

Marei, A. N., R. M. Barkhudarov, and N. Ya. Novikova. 1974. Global cesium-137 fallout and man. Moscow: Atomizdat.

[10]Protection of the public in situations of prolonged radiation exposure: ICRP publication 82. 1999. Annals of the ICRP 29(1-2). Oxford: Pergamon Press.

Gonzalez, Abel J. 2001. Decision-making about chronic radiation exposure to the public: New recommendations from the ICRP. ANRI Magazine 3(26):37-49. Available online at *www2000.irpa.net/irpa10/cdrom/01257.pdf*. Accessed August 17, 2007.

[11]Release of sites from regulatory control on termination of practices safety guide. Safety standards series No. WS-G-5.1. 2006. Vienna: International Atomic Energy Agency. Available online at *www-pub.iaea.org/MTCD/publications/PDF/Pub1244_web.pdf*. Accessed August 17, 2007.

[12]Kutkov, V. A. 2007. Evolution of the system for ensuring radiation safety in light of the new ICRP and IAEA recommendations. ANRI Magazine 1(48):2-24.

Gremikha facilities and the environmental rehabilitation of the site where they are located include the identification by the commission of three radiation exposure situations requiring various radiation safety approaches:

1. **Planned radiation exposure situations** including situations of exposure under controlled conditions involving radiation sources
2. **Accidental radiation exposure situations** including situations of uncontrolled exposure from sources that had been under regulated control but went out of control as a result of radiation accidents
3. **Existing radiation exposure situations** including situations of exposure from unregulated sources dispersed in the environment and already existing when it was decided that they needed to be controlled. These situations would include both exposure to natural radionuclides found in the environment and contamination from radiation accidents and past activities that occurred as a result of operations with sources that were not under regulated control. Examples of such situations would be exposure caused by the radiation legacy, including, among other things, exposure at contaminated sites of former military facilities located outside the regulatory system and as a result of conversion of those sites now being placed under regulatory control.

In situations of planned exposure, the ICRP uses several approaches to ensuring radiation safety, applying the concepts of dose limit, dose constraint, and reference level.

The dose limit is the level associated with a particular person that represents the maximum individual exposure from all regulated sources. This level is used for retrospective assessment of past exposure occurring within the framework of planned activities. It is also used to demonstrate that a regulated source is being operated in accordance with regulatory requirements. The main area in which this level is used is for radiation regulation in general and radiation safety status for source utilization in particular. The goal of its use is to restrict actual exposure and to provide feedback on how a device is managed and used.[13] The level of the dose limit is expressed in normative values, specifically in annual effective dose units.

The dose constraint is the level associated with a particular radiation source and represents the limit on exposure to any individual from this source. This level is used in planning source operations to limit future exposure. The main area in which this level is used is in the optimization of radiation protection overall and the creation of new sources or reconstruction of existing ones in particular. The goal of its use is to limit potential exposure and manage the source. The value

[13]Kutkov, V. A., V. V. Tkachenko, and V. P. Romantsov. 2003. Radiation safety for nuclear power plant personnel: A training handbook. 2003. V. A. Kutkov, ed. Moscow-Obninsk: Atomtekhenergo and Obninsk State Technical University for Nuclear Power Engineering, 344 pp.

BOX 21-1
System of Individual Protection Levels for Job-Related Exposure and Exposure of Members of the Public

Dose constraint and reference level: limitation on future exposure to ensure radiation safety during planning (optimization) or conduct of a regulated practice and/or exposure from a particular regulated source in any sort of planned exposure situation

Dose limit: assessment of radiation safety status based on the level of past (completed) exposure from all regulated sources in a planned exposure situation

of the dose constraint is expressed in terms of normative values, specifically in annual effective dose units.

The reference level is the value established for limiting ongoing human exposure in situations involving accidental or existing exposure as well as during planned activities. If the reference level is exceeded, certain previously planned corrective (protective) actions must be taken to reduce the risk associated with the actual exposure. The goal of its use is to ensure that the dose limit or the levels associated with the dose constraint are not exceeded. The values of the reference levels may be expressed in terms of normative and operational values as well as in terms of activity of the source or of the radioactive contamination in the environment.

The three-component structure of levels for the regulation of radiation protection presented in Box 21-1 corresponds overall with the system that has been created in Russia. Here the role of the dose constraint is played by the quota, "that part of the dose limit established to limit public exposure to a specific technogenic radiation source and means of exposure (externally, by ingestion of water or food, or by inhalation)."[14] Although the quota and the dose constraint are similar, they are not essentially equivalent.

On the whole, researchers currently see no substantial impact by the Gremikha temporary storage site on adjoining areas in the observation zone (the settlements of Ostrovnoi and Gremikha). However, the situation might change as a result of rehabilitation efforts.[15] Table 21-1 presents a list of end-state options at the temporary storage site after rehabilitation as well as strategies for achieving

[14]Radiation safety norms (RSN-99): Hygiene regulations SP-2.6.1.758-99. 1999. Moscow: Ministry of Health of the Russian Federation.

[15]Kutkov, V. A. 2007. Evolution of the system for ensuring radiation safety in light of the new ICRP and IAEA recommendations. ANRI Magazine 1(48):2-24.

them.[16] Situations related to exposure during conversion of the storage site and implementation of reprofiling strategies (S4.1, S4.2) or elimination of its facilities (S5.1, S5.2) may be categorized as three different types:

1. Planned exposure situations caused by work associated with restoration of regulated control over dangerous nuclear and radioactive materials
2. Planned exposure situations caused by the operation of the facility after its reprofiling as part of implementation of strategies S4.1 and S4.2
3. Existing exposure situations caused by remaining radioactive contamination at the location of a new radiation-hazard facility as part of implementation of strategies S4.1 and S4.2 or at the location of a new industrial facility as part of implementation of strategies S5.1 and S5.2

In accordance with the new ICRP recommendations for planning radiation protection measures for personnel during the facility conversion and rehabilitation stage, it would be expedient to establish two levels of dose constraints:

1. For planning routine work, the dose constraint could be set at a level no higher than 5 mSv per year.[17]
2. For planning extraordinary work, the dose constraint could be set at a level no higher than 25 mSv per operation.[18]

To evaluate radiation safety status, it would be appropriate to use the two dose-limit values for exposure from technogenic sources that are currently in RSN-99:

1. Personnel exposure from all sources during performance of routine work at the facilities: 20 mSv per year on average over any 5 consecutive years, but no more than 50 mSv in any single year

---

[16]Bylkin, B. K., Yu. Ye. Gorlinsky, V. A. Kutkov, O. A. Nikolsky, V. I. Pavlenko, Yu. V. Sivintsev, and B. S. Stepennov. 2007. Application of multifactor analysis in the selection of end state options and environmental rehabilitation strategies for the temporary storage site for spent nuclear fuel and radioactive waste in Gremikha village. Preprint IAE-6456/3. Moscow: Russian Research Center—Kurchatov Institute.

[17]Kutkov, V. A. 2007. Evolution of the system for ensuring radiation safety in light of the new ICRP and IAEA recommendations. ANRI Magazine 1(48):2-24.

[18]Op. cit.

Bylkin, B. K., Yu. Ye. Gorlinsky, V. A. Kutkov, O. A. Nikolsky, V. I. Pavlenko, Yu. V. Sivintsev, and B. S. Stepennov. 2007. Application of multifactor analysis in the selection of end state options and environmental rehabilitation strategies for the temporary storage site for spent nuclear fuel and radioactive waste in Gremikha village. Preprint IAE-6456/3. Moscow: Russian Research Center—Kurchatov Institute.

TABLE 21-1 Environmental Rehabilitation Strategies for Facilities at the Temporary Storage Site for Spent Nuclear Fuel and Radioactive Waste in Gremikha Village

End State	Requirements for Safety Level	Strategy
E0: Current state	Full compliance with requirements for nuclear- and radiation-hazard facilities and for placement and operation of storage or burial site for spent nuclear fuel and radioactive waste	S1: No measures taken
E1: Repository for spent nuclear fuel and radioactive waste and/or burial point for secondary radioactive waste		S2: Conversion and temporary operation of facility
		S3: Conversion and decommissioning of facility at site
E2.1: Brown field—site for radiation-hazard facility	Full compliance with requirements for radiation-hazard facilities	S4.1: Reprofiling of facility
		S4.2: Reprofiling of facility
E2.2: Brown field—site for industrial (nonradiation-hazard) facility	Exemption from regulatory control due to limits on site use	S5.1: Elimination of facility and rehabilitation of site
		S5.2: Elimination of facility and rehabilitation of site
E3: Green field	Complete exemption from regulatory control	S6.1: Elimination of facility and renovation of site
		S6.2: Elimination of facility and renovation of site

2. Public exposure from sources associated with the facility: 1 mSv per year on average over any 5 consecutive years, but no more than 5 mSv in any single year

Under the new ICRP recommendations, facility rehabilitation goals may be considered achieved if exposure doses from the remaining contamination do not exceed reference levels established by the national regulatory agency. Their values may be appropriately established depending on the end stage of the facility after the Gremikha temporary storage site has been rehabilitated:[19]

[19]Bylkin, B. K., Yu. Ye. Gorlinsky, V. A. Kutkov, O. A. Nikolsky, V. I. Pavlenko, Yu. V. Sivintsev, and B. S. Stepennov. 2007. Application of multifactor analysis in the selection of end state options and environmental rehabilitation strategies for the temporary storage site for spent nuclear fuel and radioactive waste in Gremikha village. Preprint IAE-6456/3. Moscow: Russian Research Center—Kurchatov Institute.

1. If the Gremikha storage site is reprofiled and its site and facilities are subsequently used for the creation of a radiation-hazard facility, the following reference level values are proposed for doses from the remaining contamination:

- At worksites for Group A personnel: 3 mSv per year
- At worksites for Group B personnel: 1 mSv per year
- For people living in the observation zone: 0.1 mSv per year

2. If the strategy for eliminating the Gremikha storage site is implemented (S5.1 and S5.2) and the site is subsequently used for a general industrial facility, the following reference level values are proposed for doses from the remaining contamination:

- For personnel at the industrial facility: 1 mSv per year (0.9 mSv per year as a result of work at the site and 0.1 mSv per year due to living in the observation zone)
- For people living in the observation zone: 0.1 mSv per year

3. If the option for further use of the Gremikha storage site as a radiation-hazard facility is implemented, the design of such a facility and its operations must ensure that annual exposure doses resulting from the activities of the facility do not exceed established dose constraint levels:

- For Group A personnel: 7 mSv per year
- For Group B personnel: 1 mSv per year
- For people living in the observation zone: 0.15 mSv per year

During the stage of planning work and designing radiation protection measures, a final decision must be made on the values of the above-mentioned regulatory levels, with interested parties to be involved in the decision-making process in accordance with the new ICRP recommendations.

22

Cleaning Up Sites Contaminated with Radioactive Materials: Coastal Maintenance Bases Andreev Bay and Gremikha

Dieter K. Rudolph, Science Applications International Corporation

INTRODUCTION

During the Cold War the Soviet Union built 248 nuclear-powered submarines. Very few older submarines were retired as more capable ones joined the fleet. About two-thirds of the submarines were assigned to the Northern Fleet in northwestern Russia. This area covers Murmansk Oblast, where Andreev Bay and Gremikha are located on the Kola Peninsula, and Arkhangelsk Oblast, where Zvezdochka, one of the nuclear submarine dismantlement shipyards, is located in Severodvinsk.

The coastal maintenance bases at Andreev Bay and Gremikha began operating in the early 1960s to support Northern Fleet nuclear submarine operations and to manage spent nuclear fuel and radioactive waste. Andreev Bay serviced only submarines with pressurized water reactors and operated until 1985. Gremikha also serviced the liquid-metal coolant reactors for the Alpha-class nuclear submarines based there. Operations at Gremikha ceased in 1992. After years of neglect, responsibility for the coastal maintenance bases was transferred from the Russian Navy to the Ministry of Atomic Energy (Minatom) in 2000. The special Northern Federal Enterprise for Radioactive Waste Management (SevRAO) was established within Minatom (now Rosatom [Russian Federal Atomic Energy Agency]) to provide administrative oversight and management of the facilities. The coastal maintenance bases were renamed as spent nuclear fuel and radioactive waste temporary storage facilities—TSFA (Andreev Bay) and TSFG (Gremikha)—to reflect their current status.

Mass Decommissioning

After the collapse of the Soviet Union, severe reductions in defense spending left the Russian Federation Navy unable to maintain a large, active submarine fleet and its supporting infrastructure, including the coastal maintenance bases. Routine facility maintenance of coastal maintenance bases ceased. In 1994, only 35 percent of the funds earmarked for the Russian Northern Fleet were actually transferred.[1] The lack of funds and the fact that many of the first- and second-generation nuclear submarines exceeded their service life led to mass decommissioning and the neglect of the supporting infrastructure, including the coastal maintenance bases. The large influx of spent nuclear fuel, radioactive waste, and toxic waste associated with decommissioning and dismantling nuclear submarines overwhelmed an already burdened system, resulting in severe problems of safe management of spent nuclear fuel, radioactive waste, and toxic waste.

Another factor in reducing the fleet size was the Strategic Arms Reduction Treaty, which required the elimination of more than 40 ballistic submarines with more than 600 ballistic missile launchers.[2] Although it increased the number of submarines to be dismantled, it had a positive effect because the United States funded the modernization of dismantlement facilities at a number of shipyards, including Zvezdochka in the northwest and in Russia's Far East Region through the Cooperative Threat Reduction Program. This program eliminated bottlenecks in spent nuclear fuel and radioactive waste management and until recently paid for the dismantlement of the entire ballistic submarines. Now, funding is limited to removal of the launcher tubes and reactor compartments. The facilities provided through the Cooperative Threat Reduction Program are available for Russian "general purpose" (nonballistic nuclear submarines) dismantlement on a not-to-interfere basis.

As of February 2007, a total of 198 nuclear submarines were decommissioned and 148 were dismantled. In the Northern Fleet, 120 nuclear submarines were decommissioned and 97 are dismantled. Ten Northern Fleet submarines are in the process of being dismantled.[3]

---

[1]Nikitin, A., I. Kudrik, and T. Nilsen. 1996. The Russian Northern Fleet: Sources of Radioactive Contamination. Bellona Foundation: Oslo, p. 15. Available online at *www.bellona.org/reports/The_Russian_Northern_Fleet*.

[2]Federation of American Scientists: *www.fas.org/nuke/control/start2/*.

[3]Akhunov, V. 2007. Progress on Dismantlement of Nuclear Submarines under the Global Partnership International Cooperation. Presented at the Northern Dimensions Environmental Partnership-Nuclear Operating Committee/Contact Experts Group Workshop on Results of Strategic Master Plan, Phase 2, London, April 10, 2007. Available online at *www.iaea.org/OurWork/ST/NE/NEFW/CEG/documents/ws042007_1E.pdf*.

Spent Nuclear Fuel Management

Russia uses a "closed fuel cycle" that requires reprocessing the spent nuclear fuel at the Mayak Chemical Complex to recover uranium. Spent nuclear fuel is transported in special containers that can hold up to 49 fuel assemblies. Ten to 12 containers are needed to defuel one submarine, since the typical Russian nuclear submarine has two reactors, and each reactor holds about 248-252 fuel assemblies. The spent nuclear fuel containers are transported by technical support vessels from Andreev Bay and Gremikha to two facilities in northwest Russia: RTP Atomflot in Murmansk or Zvezdochka Shipyard in Arkhangelsk. The containers can be loaded directly onto special railcars or placed in storage until the railcar arrives. One train typically transports 12 containers—the fuel from one submarine—for a distance of more than 2,000 km to Mayak. Certain fuel cannot be reprocessed at Mayak. This includes fuel from liquid metal coolant reactors, damaged fuel stored at Andreev Bay and Gremikha, fuel stored on the *Lepse* (a technical support vessel that is the biggest nuclear and radiation risk of all retired nuclear service ships in Russia), and zirconium-clad fuel from nuclear icebreakers. Special provisions must be made for the management of this fuel.

SPENT NUCLEAR FUEL AND RADIOACTIVE WASTE TEMPORARY STORAGE FACILITY ANDREEV BAY

Andreev Bay is one of four naval facilities of the Zapadnaya Litsa Naval Base, the largest Northern Fleet base that maintains most of the active nuclear submarines. Andreev Bay, located along the western side of the Litsa Fjord, is the only facility where no submarines are based. It is strictly used as a storage facility for spent fuel and radioactive waste and is only 45 km from the Norwegian border. The other three facilities—Bolshaya Lopatka, Malaya Lopatka, and Nerpichya—are situated along the eastern side of the fjord. The Typhoon-class ballistic submarines are based at Nerpicha, the only facility with a rail connection. The military families live in Zaozersk, a closed administrative town with about 30,000 inhabitants.

Andreev Bay has the largest inventory of spent nuclear fuel and radioactive waste: 22,000 spent fuel assemblies, 18,000 m^3 of solid radioactive waste, and 3,400 m^3 of liquid radioactive waste. Routine maintenance at the base stopped in 1993, and with the harsh Arctic conditions it did not take long for most of the infrastructure to be destroyed, including heat, water, power supply, and sewage.[4] In October 2001 the U.S. Department of Energy hosted an International Atomic Energy Agency (IAEA) contact expert group workshop at Idaho Falls, Idaho, to consider international assistance on reducing contamination and safety risks as-

[4]Vasiliev, A. P. 2009. The past, present, and future of the facilities at Andreev Bay. Pp. 127-136 in Cleaning Up Sites Contaminated with Radioactive Materials: International Workshop Proceedings. Washington, D.C.: The National Academies Press.

sociated with spent nuclear fuel and radioactive waste management at Andreev Bay. This served as a catalyst for international cooperative efforts that include Norway, the United Kingdom, and Sweden.

Spent Nuclear Fuel

The Soviet spent nuclear fuel management policy in effect when Andreev Bay began operating in the early 1960s required wet storage in cooling pools for 5 to 7 years before the fuel was sent for reprocessing at Mayak. The spent nuclear fuel, housed in special containers holding five to six spent fuel assemblies, was unloaded from technical support vessels onto trucks for further transport to Building 5, where the containers were suspended by chains into two small cooling pools. By 1973, additional spent nuclear fuel storage was required and an annex with two larger cooling pools was constructed. In 1982 a leak in one of the large cooling pools was discovered. Efforts to stop the leak by pouring concrete into the pool and under the building foundation were not successful. Leakage was also detected in the second large pool. Some of the spent nuclear fuel canisters fell to the bottom of the pool after restraining chains broke, and cesium-137 and strontium-90 penetrated into the ground beneath the building with the leaking cooling water contaminating a creek flowing under the building.[5] The walls of the building and the foundation were saturated with radioactive substances and have a dose equivalent rate between 9,000-20,000 μSv per hour. Efforts are under way to determine the extent and depth of soil contamination along the foundation. Some of the ground between Building 5 and the coastline has a dose equivalent rate of 450 μSv per hour. The sediment at the bottom of the cooling pools may contain spent nuclear fuel fragments.

As an emergency measure, the Soviet Navy decided to reconfigure three large, 1,000 m^3 storage tanks for temporary (5-7 years) dry storage of the spent nuclear fuel from Building 5. The tanks were designed to hold liquid radioactive waste but were not used for this purpose. They were modified by installing steel cylinders 25-27 cm in diameter, about 1,000 cylinders per tank, and then filling the space between the cylinders with concrete. A container holding up to seven fuel assemblies was then inserted into each cylinder. Dry storage units (DSUs) 3A, 2A, and 2B were placed in operation as they were completed in 1983, 1985, and 1986, respectively, and the transfer of spent nuclear fuel from Building 5 was completed in 1989. The 6-year service life of the dry storage units expired long ago. Groundwater has penetrated through the floors, the roofs leak, and many of the spent nuclear fuel cells contain water. A comprehensive engineering and radiological survey revealed that cells in DSU 2B and 3A are filled with high

[5]Vasiliev, A. P., V. P. Vasoukhno, M. E. Netecha, et al. 2006. Radiological Condition of Andreeva Bay Territory and Water Area. Universal Decimal Classification number 621.039.76+614.876. Available online at *www.bellona.org/filearchive/fil_AndreevaBayRadiationSurvey_eng.pdf.*

salt content water, including chlorides, which is degrading the spent nuclear fuel. Calculations by Russian and British experts using "conservative assumptions" indicate a spontaneous chain reaction may occur during retrieval if the bottom of a canister containing the fuel breaks off during lifting.[6]

Spent nuclear fuel is also stored on an open pad near the DSUs. More than 360 fuel assemblies are stored in old TK-6 and TK-11 containers. Most of the spent nuclear fuel is damaged. Special provisions are required for the management of damaged spent fuel, since it cannot be reprocessed at Mayak. One of the DSUs contains icebreaker fuel that has zirconium cladding, and this fuel also cannot be reprocessed at Mayak.

In summary, an accurate inventory of spent nuclear fuel does not exist, the condition of the fuel is unknown, and therefore procedures for safe management have not been developed. Facilities for repackaging this spent nuclear fuel do not exist. The final disposition of the fuel has not been decided, since there is much damaged fuel that cannot be reprocessed at Mayak.

Solid Radioactive Waste

Andreev Bay has the largest accumulation of solid radioactive waste, and it is stored on three open pads and in seven storage vaults. The solid radioactive waste volume was originally estimated at 6,000 m^3, but based on recent surveys this was increased to about 18,000 m^3. Some of the storage areas contain a mix of high-, medium-, and low-level solid radioactive waste, and the survey of subsurface facilities has not been completed. There is no facility for processing solid radioactive waste, and additional radioactive waste will be generated as spent nuclear fuel and radioactive waste management activities are implemented.

Liquid Radioactive Waste

Andreev Bay has the largest volume of liquid radioactive waste—more than 3,000 m^3. The liquid radioactive waste management facility was constructed, but the equipment was not installed. The centralized collection system for liquid radioactive waste has been destroyed. The service life of all liquid radioactive waste storage tanks has expired. The leaks in spent nuclear fuel DSUs 2B and 3A and in subsurface solid radioactive waste facilities continue to generate additional liquid radioactive waste. About 60 metric tons of water per year seep through the solid radioactive waste storage pad foundation into the basement of Building 6. About 300 metric tons of liquid radioactive waste has accumulated in the "dry" storage units, where spent nuclear fuel is stored. Liquid radioactive waste is also found in the solid radioactive waste storage vaults, again due to leaks. There is no facility for liquid radioactive waste treatment other than RTP Atomflot.

---

[6]Vasiliev, op. cit.

Grounds and Aquatic Area

Building 5 and the solid radioactive waste storage pad are sources of radioactive contamination for the territory. A large area between Building 5 and the coastline is contaminated with radionuclides at a depth of 1 m. The creek water contamination levels from Building 5 remain unchanged since 1982. This supports the assumption that the leaks from the cooling pools were major and that radionuclides penetrated the soil and underground formations. Areas around the solid radioactive waste storage pad indicate contamination of the soil is substantially higher than permitted. Aquatic surveys were conducted in 1997 and 1999. Several hot spots were identified where the creek enters the bay areas and along the coastline closest to the solid radioactive waste storage pad. No surveys have been conducted since 1999.

Ongoing Efforts

The infrastructure has been improved to address the most pressing problems. Facilities for workers have been installed, the road has been repaired, and a radiation-monitoring system is being installed. This is an international effort, with Norway taking the lead on infrastructure improvements and conducting initial site and radiation surveys. Sweden is responsible for surveys of liquid radioactive waste and the open solid radioactive waste storage sites. The United Kingdom has the lead on a comprehensive engineering and radiation survey of spent nuclear fuel in the DSUs and in Building 5. DSUs 2B and 3A require special procedures for spent nuclear fuel removal.

Workshop Discussions

During the workshop, the following questions and concerns were raised and addressed:

- A radiation-monitoring system tied into a central monitoring site will monitor all hazardous areas, which will keep workers safe during all stages of spent nuclear fuel and radioactive waste removal.
- Spent nuclear fuel management is a priority issue. Handling procedures for the old TK-6 and TK-11 spent nuclear fuel casks and canisters with spent nuclear fuel from the cells of the dry storage unit are developed. Procedures for handling and processing damaged fuel have been developed. The transport of spent nuclear fuel is planned by using a vessel that will be provided by Italy. There is a plan to construct a spent nuclear fuel transfer pad at Andreev Bay so that the fuel can be sent to Mayak via Zvezdochka shipyard in Severodvinsk, where there is an onshore defueling facility constructed by the U.S. Cooperative Threat Reduction Program. There was also mention of the possibility of building

a new bridge solely for the transport of spent nuclear fuel, as the existing bridge is old and must be inspected before each spent fuel shipment.

- There are plans to process liquid and solid radioactive wastes onsite so that the backlog does not increase.
- Creation of a special category of very low-level waste (0.3-100 Bq/g) is being considered. The bulk of the radioactive waste at Andreev Bay—namely soil and structural components—falls into this category. France and Sweden are using this category, as it allows for simplified cost-saving disposal procedures.

The following additional points were made during the presentation and are also applicable to Gremikha:

- Consider mobile containerized systems for solid and liquid radioactive waste management for ease of transport and possible future reuse. A mobile solid radioactive waste treatment facility and a modular storage complex were installed by the Arctic Military Environmental Cooperation Program at Shipyard 10, Polyarny.
- Accurate characterization of the waste stream is essential for proper waste processing.

SPENT NUCLEAR FUEL AND RADIOACTIVE WASTE TEMPORARY STORAGE FACILITY GREMIKHA

Gremikha is located about 35 km east of the mouth of the Murmansk Fjord and is accessible only by sea or helicopter. Climatic conditions are severe, with abrupt temperature variations. The strong humid winds with speeds greater than 15 m per second gave rise to the nickname "land of the flying dogs." Winter storms can keep Gremikha isolated for extended periods. Ostrovnoi, a closed administrative town, provides the workforce for the base. The population is shrinking, down from 10,000 in 2004 to only 3,000 inhabitants in 2007. Routine transport of supplies and people is provided by the support vessel *Klavdia Yelanskaya*. Cargo cannot exceed 2 metric tons, because of weight limitations. Murmansk and Arkhangelsk are equidistant at 400 km, and therefore both RTP Atomflot and Zvezdochka are viable options as spent nuclear fuel transshipment sites.

As a coastal maintenance base, Gremikha was the home base for the Alpha-class submarines, known for their high speed, titanium hulls, and liquid-metal coolant reactors with fuel enrichment up to 90 percent. Refueling Alpha-class submarines requires removal and replacement of the reactor core. This process requires special facilities, including a dry dock (Building 2) specifically modified for this purpose, a reactor refueling building (Building 1A), a spent removable core storage facility (Building 1B), and a boiler house to heat the lead-bismuth alloy to greater than 125°C to keep it liquid. The base managed the refueling of both liquid-metal coolant and pressurized water reactors until its operations ceased in 1992. Most of the infrastructure is now degraded and much of the equipment is inoperable.

Spent Removable Cores and Spent Nuclear Fuel

Constructed in 1989, Building 1A was designed for the unloading and temporary storage of spent removable cores from liquid-metal coolant reactors of the Alpha-class submarine. The dry dock and handling equipment for removing spent removable cores has been restored and is in working condition. One core was removed in 2005 and another was removed in 2006. Eight cores are currently in special storage facilities. The service life of the storage facilities expires in 2008, but steps are under way to improve the facilities and upgrade the monitoring system and subcriticality control system so that its service life can be extended to 2015.

There is a problem with long-term storage of spent removable cores. The original design presumed that after several years of cooling the cores would be transported to the reprocessing plant because of their relatively small subcriticality. Studies have shown that during "freezing" of the spent removable core, pores and cavities can form in the eutectic, and criticality can be reached with the ingress of 20 kg of water.

The spent nuclear fuel and radioactive waste are stored in nonstandard conditions. Protective barriers of many spent fuel assemblies and solid and liquid radioactive waste containers leak, so further environmental contamination is evident. The infrastructure does not provide nuclear and radiation safety protection, especially for liquid-metal coolant fuel assemblies and spent reactor cores.

For more than 30 years, 116 old TK-6 and TK-11 spent nuclear fuel containers have been stored on an uncovered outdoor pad mixed with solid radioactive waste. Many of the containers have leaking lids, and water has entered some of the containers. It is estimated that almost 800 spent fuel assemblies are in the containers on the pad. A detailed inventory of the containers and fuel assemblies has started. The pad poses a real threat to the environment and contaminates the territory.[7] Based on preliminary findings, it is estimated that more than 30 percent of the fuel is damaged.

Building 1 has four ferroconcrete cooling pools for the storage of spent fuel assemblies from pressurized water reactors. Pool 1 began leaking in 1986 and is one of the main sources of radioactive contamination of the site.[8] Spent fuel

[7]IAEA Contact Expert Group Database, Project 45, FR2: Gremikha Site: Feasibility study for the rehabilitation and urgent actions. Available online at *cegdb.iaea.org/ProjectDetails.aspx?ProjID=45.*

[8]Kovalenko, V. N., and V. A. Mazokin. 2003. Situation at the Gremikha Base: Main Problems and General Plan for Remediation. Paper presented at the IAEA Contact Experts Group Workshop in Cadarache, France, October 29-31, 2003. Available online at *www.iaea.org/OurWork/ST/NE/NEFW/CEG/documents/ws102003_kovalenko-e.pdf.*

assemblies are stored in 106 old canisters (shrouds) in three of the pools. It is estimated that more than two-thirds of the fuel is damaged and requires special handling because of bending, swelling, and breaks in the fuel rods. Special equipment and procedures will be required for the safe handling and transport of the fuel.

Solid Radioactive Waste

About 800 m^3 of mixed solid radioactive waste (high-, intermediate-, and low-level) is stored in a variety of places, including Building 19, the sumps of the decontamination pad near Building 1, and the open solid radioactive waste storage pad. The Integrated Engineering and Radiation Survey conducted in 2006 identified high-level solid radioactive waste in Sumps 1 and 2 near Building 1 and in concrete containers on the open storage pad. The plan is to treat the high-level waste in the sumps as a separate project because special devices for removal into shielded containers will be required.

Additional solid radioactive waste will be generated as work proceeds with spent nuclear fuel and radioactive waste management processes, decontamination of equipment, and rehabilitation of buildings, grounds, and aquatic areas. About 1,500 m^3 of solid radioactive waste will be generated as part of the remediation efforts.

Liquid Radioactive Waste

Plans for a liquid radioactive waste treatment plant and a cementation facility for liquid radioactive waste never materialized. A number of underground tanks are used for storage of liquid radioactive waste as well as floating tanks. The total volume of liquid radioactive waste is about 150 m^3. High- and intermediate-level liquid radioactive waste is found in the spent fuel containers on the open solid radioactive waste storage pad. It is expected that an additional 300 m^3 of waste will be generated as part of the remediation efforts.

Grounds and Aquatic Area

The open solid radioactive waste storage pad is the most contaminated area at Gremikha. It sits on a hill, and the rain and snowmelt wash contaminants into the grounds and the aquatic area. Many of the spent fuel and radioactive waste containers as well as the liquid radioactive waste tanks are degraded and contaminate the grounds and aquatic area. There is a trend towards further contamination of the environment.[9]

[9]Strategic Master Plan Decommissioning of Retired Nuclear Fleet and Environmental Rehabilitation of its Supporting Infrastructure in Northwest Russia, Part 1. Characteristics of the Facilities Subject to Decommissioning (Rehabilitation), Goals and Determination of Priority Project Lists.

Ongoing Efforts

Radiation surveys were conducted in 1999 and 2003. The following excerpt from a June 23, 2003, letter from V. D. Safutin, director of the Russian Research and Design Institute of Power Technology (VNIPIET), to SevRAO and the International Center for Environmental Safety summarizes what was accomplished during this period: ". . . no real work aimed at improving conditions at the coastal technical base in Gremikha was performed in the period. The spent fuel assemblies within casks TK 6 and TK 11 and the radioactive waste storage facilities continue to deteriorate, thus impairing the environment. Practically the entire infrastructure has been destroyed, making the previously developed technological approaches to spent nuclear fuel and radioactive waste management of little use . . . ."[10]

International attention was not focused on the problems at Gremikha until October 2003, when the IAEA Contact Experts Group organized a special workshop at Cadarache, France, to stimulate international cooperation. The workshop participants identified priority tasks and feasibility studies for the management of spent nuclear fuel from pressurized water reactors and liquid-metal coolant reactors and management of radioactive waste, as well as projects at the site. The 12 highest priority projects are focused on ensuring the radiation safety of workers, restoring the facility's basic infrastructure, and completing an integrated engineering and radiation survey of spent nuclear fuel and radioactive waste. A detailed survey of spent nuclear fuel on the solid radioactive waste pad and in Building 1 is ongoing. The fuel assemblies from pressurized water reactors require special procedures and equipment for handling and transport. Because the containers have leaking lids and some lids cannot be opened, they must be repackaged before transport. Italy plans to design and provide new containers. Issues regarding management and reprocessing of spent removable cores from liquid coolant reactors and their reprocessing technology have not been resolved, and this is an urgent issue. Efforts are under way to extend the service life of the reactor core storage facility to 2015 by improving the monitoring system. International partners include France, the European Bank of Reconstruction and Development, the European Commission (Technical Aid to the Commonwealth of Independent States Program), and Italy.

Workshop Discussions

The following questions and concerns were raised and addressed:

---

[10]Vasiliev, A. P. 2003. Projects Proposals for Radioactive Waste Management, Remediation of Buildings, Structures and Territory of Temporary SNF and RW Storage Facility in Gremikha. Paper presented at the IAEA Contact Experts Group Workshop in Cadarache, France, October 29-31, 2003. Available online at *www.iaea.or.at/OurWork/ST/NE/NEFW/CEG/documents/ws102003_vasiliev-e.pdf*.

- Environmental remediation of Gremikha is a complex problem that will take time. The strategy is to remediate the site stage by stage, making incremental improvements in the security and safety of the facility. This includes providing secure and safe intermediate storage of spent nuclear fuel and radioactive waste at Gremikha; the final storage site will be determined later. Existing facilities will be used for secondary radioactive waste storage once they are secured and until the final end state of remediation at Gremikha is determined.
- It is estimated that rehabilitation of Gremikha will cost 187 million euros and can be accomplished in 12 to 20 years, depending on the final decision as to the future use of Gremikha. The work will be done in four stages:

1. Conversion: complex engineering and radiation examination, maintenance, and secure and safe storage of spent nuclear fuel and radioactive waste
2. Temporal operation: removal of spent nuclear fuel and radioactive waste from the site
3. Decommissioning: dismantlement of facilities and buildings
4. Cleanup: decontamination of facilities, grounds, and aquatic areas

- A special presentation on criteria of ecological rehabilitation of Gremikha was provided. A number of options including "brown field" and "green field" status were investigated. Multiattribute decision analysis was used to identify a solution. Ensuing discussions pointed out that this approach should be used as a tool in finding a solution, but not to identify a specific solution.
- Italy will provide special handling equipment for the old TK-6 and TK-11 spent nuclear fuel containers. They will also build or provide a ship for transport of the containers. The *Lotte* is also considered for transport if the Italian ship does not arrive in time.
- A number of sites with special "hot chambers" have been considered for defueling the spent removable cores, including Obninsk and Dimitrovgrad, and the latter will probably be the site that is selected.
- The shrinking workforce (from 10,000 to 3,000) is a problem. A final decision on the special status of the town has not been made.
- Measures to extend the service life of the reactor core storage facility are in progress, but the presence of europium in the spent removable cores presents a problem that remains unresolved.
- There was also mention of building a new bridge just for the transport of spent nuclear fuel, as the existing bridge is old and must be inspected before each spent nuclear fuel shipment.

STRATEGIC MASTER PLAN

SMP-1: The scope and complexity of dismantling the nuclear vessels taken out of service and restoring the support bases into environmentally acceptable condition requires an overall strategy to guide Russia and its international partners in defining and prioritizing projects. The Northern Dimension Environmental Partnership (the first phase of the Strategic Master Plan SMP-1) was completed in 2004. This phase of the plan provides a detailed analysis and a conceptual strategy for addressing the Northern Fleet nuclear legacy issues of dismantling nuclear-powered vessels (submarines and surface ships), along with their reactor units and technical support vessels, including those operated by RTP Atomflot, and cleaning up coastal maintenance bases (Andreev Bay and Gremikha). SMP-1 identified, but did not rank, 21 high-priority and 24 priority measures requiring critical attention. Two-thirds of the high-priority measures pertain to Andreev Bay or Gremikha or both, as shown in Table 22-1.

Strategic Environmental Assessment (SEA)

SMP-1 was evaluated by the European Bank of Reconstruction and Development to assess the environmental impact and risks of planned activities to the population. The strategic environmental assessment (SEA) was completed in 2005. Considering the accidental release of radioactivity into the environment the main threat, the SEA further analyzed the SMP high-priority and priority measures as follows:

- The highest overall priority is the refurbishment of the Gremikha reactor storage facility and the subsequent decommissioning of the reactor cores from the defueled Alpha-class submarines "due to the high risk of a nuclear accident that would result if a relatively small amount of water were to leak into the storage facility." In a footnote, the authors of the SEA explain that "recent Russian studies found that the ingress of water into the cores of Alpha-class submarines is not feasible. If confirmed, then addressing the state of storage of these cores

TABLE 22-1 High-Priority Measures from the Strategic Master Plan

Measure	Andreev Bay	Gremikha
Safe management of spent nuclear fuel	X	X
Safe management of spent removable cores		X
Integrated engineering and radiation survey of buildings, structures, grounds, and aquatic area, including an inventory of spent nuclear fuel and radioactive waste	X	X
Restoration of infrastructure for spent nuclear fuel management	X	X
Restoration of infrastructure for spent removable core management		X
Reconstruction of the spent removable core storage facility		X
Provision of physical protection of the site	X	X
Provision of radiation safety of personnel	X	X
Elimination of open-air storage pad		X

would be a lower priority." During the workshop, these recent studies were not mentioned.

- The second-highest priority is the decommissioning of the spent nuclear fuel storage facilities at Andreev Bay and Gremikha "due to the poor storage conditions, the high radionuclide inventory, and leakage of radioactivity into the environment, and the risk of nuclear accidents at these facilities."[11]

An important aspect of the SEA was the scoping consultations with the public, government and regional authorities, experts, industry, and nongovernment organizations to inform them about the SMP and to listen to concerns and suggestions. The public asked to be kept informed and that the proceedings should be transparent.

SMP-2: The second phase of the SMP is planned for completion by August 2007. The goal is to provide a fully justified integrated program for the decommissioning and rehabilitation of Navy facilities in northwestern Russia. A key part of the integrated approach is the Priority Project Program, which addresses the high-priority and priority measures identified in the first phase of the SMP. Priority projects for Andreev Bay include construction of spent nuclear fuel management facilities to prepare for spent nuclear fuel removal and solid radioactive waste management facilities for long-term storage and subsequent removal. Priority projects for Gremikha include ensuring safe storage of Alpha-class spent removable cores and removal of spent nuclear fuel and radioactive waste from open pads.

It is important to note that the SMP only addresses naval nuclear legacy issues in northwestern Russia. It does not address nuclear legacy issues in Russia's Pacific Fleet, where some experts believe the problems are actually more severe, and it does not include an analysis of civilian spent fuel, which provides the primary input for reprocessing spent fuel at Mayak.

CONCLUSIONS

1. The problems at Andreev Bay and Gremikha are similar, only the scale of the problem is smaller at Gremikha. The Andreev Bay Coordination Group and the Gremikha Coordination Group must be closely linked so that lessons learned at one site may be applied to the other.

2. Existing regulations governing the cleanup of radioactive contaminated sites seem to be ineffective, and more specific policy is required. Laws and regulations address normal conditions and are not adequate in addressing the exceptionally unsatisfactory circumstances found at Andreev Bay and Gremikha.

---

[11]Strategic Environmental Assessment Report for Strategic Master Plan of Northern Dimension Environmental Partnership (Nuclear Window), NNC Document Registry Info: 12124/TR/002, April 2005.

3. To date, international efforts have focused primarily on submarine dismantlement and not the most urgent problems, namely spent reactor core and damaged fuel management and storage. It is important to move from the feasibility study phase to tangible work as quickly as possible.

4. A long-term solution must be found for spent fuel that cannot be reprocessed at Mayak. This includes fuel from liquid-metal coolant reactors, damaged fuel stored at Andreev Bay and Gremikha, fuel stored on the *Lepse*, and zirconium-clad fuel from nuclear icebreakers. An interim storage facility has been constructed at RTP Atomflot in Murmansk for temporary storage of this fuel, but a long-term solution is needed.

5. Consultations held during the SEA indicated that the public wants to be kept informed in nontechnical terms on the progress of cleanup efforts. Keep the public informed and keep the proceedings transparent as the SMP is implemented.

6. The general requirements of how agencies will work together are unclear. For example, certification of the spent nuclear fuel interim storage and transfer pad built by the Arctic Military Environmental Cooperation at RTP Atomflot in Murmansk was delayed due to Russian agency jurisdictional disputes. The pad uses the Murmansk Shipping Company crane to lift the spent nuclear fuel casks. The Murmansk Shipping Company reports to the Ministry of Transportation. RTP Atomflot, which operates the pad, reports to Rosatom, and is regulated by the Russian Federal Inspectorate for Nuclear and Radiation Safety (Rostekhnadzor). The spent fuel is naval fuel, and the Ministry of Defense has responsibility, with regulatory oversight provided by the military inspectorate for nuclear and radiation safety. During a meeting with Rostekhnadzor in June 2007, it was confirmed that this jurisdictional problem still exists. This problem needs to be resolved.

7. Donor countries should not fund projects that will increase Russian military capabilities. The dual-use (civilian-military) issue was a factor in Japan's decision not to fund a railroad link at Zvezda, the Russian dismantlement shipyard for the Pacific Fleet, since it could be used for refueling active-duty submarines. The dual-use consideration should also be applied by foreign donors to the request for funding to repair a bridge across the Severnaya Dvina River, which is used by trains transporting nuclear fuel, both spent and fresh.

8. SMP strategy is too narrowly focused on only Northern Fleet spent nuclear fuel and radioactive waste. It does not address Pacific Fleet issues, and more importantly, in trying to identify bottlenecks, omits consideration of civilian sources of spent nuclear fuel and radioactive waste and the impact this has on Mayak.

9. The end state of Gremikha and Andreev Bay has not been decided. This decision needs to be made soon, as it will determine the extent of work required.

10. The workforce at Gremikha is declining. In the late 1980s, when

Gremikha was an active base, the population of Ostrovnoi was 30,000. As the base closed and operations ceased, the town's population decreased to 10,000 by 1998[12] and is now 3,000. The status of Ostrovnoi as a "closed military town" is about to change, and the special compensations provided by the government to keep the workforce in place will end. A continued decline in the population and thus the workforce is expected. This will have a significant impact on implementation of projects to clean up and rehabilitate the site.

[12]Kudrik, Igor. 1998. Gremikha to grip share in subs decommissioning. Brief paper posted online at *www.bellona.org/english_import_area/international/russia/navy/northern_fleet/decommissioning/7692*. Oslo: Bellona Foundation.

Other Contributions

23

Criteria for Categorizing Territories at Russian Federal Atomic Energy Agency Enterprises Experiencing Chemical and Radioactive Contamination*

S. N. Brykin, All-Russian Scientific Research Institute of Chemical Technology
N. K. Shandala, State Science Center Institute of Biophysics
N. S. Roznova, All-Russian Scientific Research Institute of Chemical Technology
A. V. Titov, State Science Center Institute of Biophysics

Two circumstances led us to develop the following criteria to categorize territories at Russian Federal Atomic Energy Agency (Rosatom) enterprises exhibiting chemical and radioactive contamination.[1] The first is the presence in Russia of sites that continue to be affected by technogenic contamination due to radiation accidents and previous activities involving nuclear technologies. The second is the need to remove from operation the facilities still presenting radiation hazards that have outlived their usefulness as well as the possibility of reducing the size of areas devoted to protected sanitary zones or observation zones around enterprises.[2] In these instances, matters related to the future use of the lands thus freed up must be addressed.

From the standpoint of ensuring environmental security, the most substantial issues involve the consequences of defense activities that polluted the environment in the first years when nuclear weapons were produced. These activities

*Translated from the Russian by Kelly Robbins.

[1]The April 1986 Chernobyl accident was a special case, and contamination resulting from it is not addressed in this paper.

[2]Sanitary zones are established in areas immediately surrounding enterprises or sites, and access to them is stringently controlled. Observation zones are farther removed from the enterprise or site but have monitoring systems in place to collect relevant environmental data.

led to the accumulation of large volumes of radioactive waste that have not been placed in environmentally safe conditions.[3]

Although today it is recognized that the normal operations of modern nuclear power plants and other nuclear industry facilities are not associated with any significant risks to public health or the environment, in the foreseeable future the industry must address a wide range of resource-consuming environmental problems. One of these broad issues, the rehabilitation of environmental media, must be resolved in order to ensure long-term environmental security.[4]

SCOPE OF ENVIRONMENTAL REHABILITATION

In late 2005 the total contaminated area associated with the nuclear industry was 474.6 km^2, including the following:

- Industrial sites—62.6 km^2
- Protected sanitary zones—215.1 km^2
- Observation zones—196.9 km^2

A total of 26 organizations in the Russian nuclear industry and nuclear power sector had sites contaminated with radionuclides, including 16 organizations belonging to Rosatom. The organizations with contaminated sites are located in 20 regions (7 different federal districts) of the Russian Federation. The breakdown of contaminated territories by location zones—on industrial sites, in protected sanitary zones, and in observation zones—is illustrated in Figure 23-1.[5]

Contamination of sites at nuclear industry organizations and power facilities is defined mainly in terms of the nuclides cesium-137, strontium-90, and plutonium-239, as well as nuclides of uranium, thorium, and radium.

According to currently existing industry recommendations, the decision to rehabilitate radionuclide-contaminated territories at industrial sites and in protected sanitary zones must be made if the open-air dose exceeds dose limit B for personnel or if the dose level under external full-body exposure totals 2.5 μGr per hour.[6] Figure 23-2 illustrates dose levels of gamma radiation at radiation-

[3]Solonin, M. I., ed. 2003. Industry safety report of the Russian Federation Ministry of Atomic Energy. Moscow: Komtekhprint Publishing House, 124 pp.

[4]Op. cit.

Brykin, S. I., and I. S. Serebryakov. 2005. Recommendations in the 2005 industry safety report: Materials from the Industry Department of Environmental Protection of the Federal State Unitary Enterprise—All-Russian Scientific Research Institute of Chemical Technology, 49 pp.

[5]Op. cit.

[6]The dose limit for category B, 5,000 μSv, is divided by the number of working hours in a year, 2,000, to equal 2.5 μGr per hour. Criteria for decision-making on rehabilitation of sites as a result of the operations of nuclear industry enterprises. 1997. Moscow: All-Union Scientific Research Institute of Solid Fuel Chemistry, 28 pp.

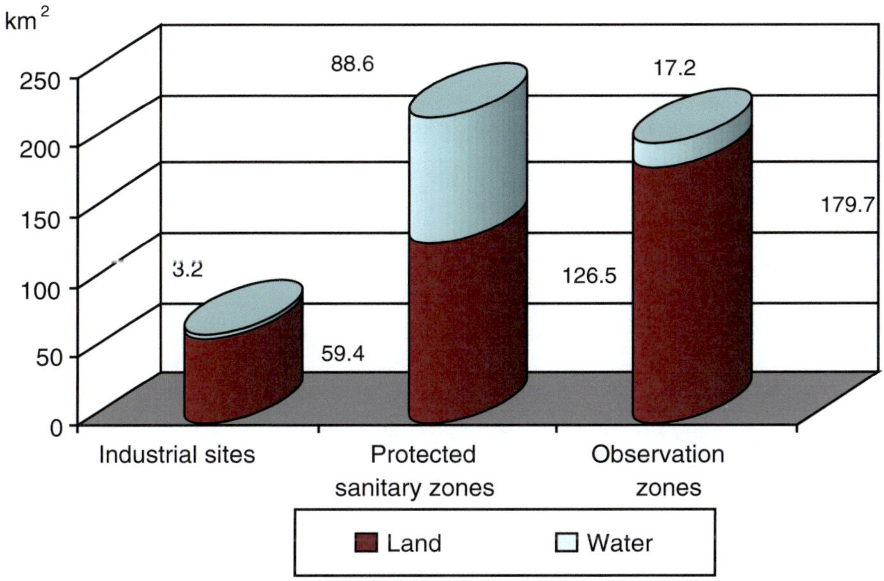

FIGURE 23-1 Distribution of radionuclide-contaminated territories by location zones.

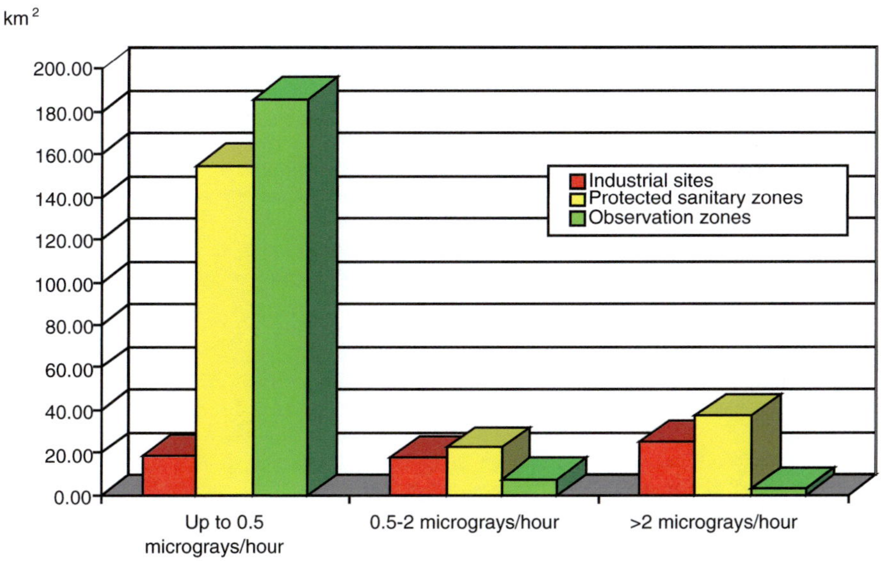

FIGURE 23-2 Distribution of radiation-contaminated sites by dose levels of gamma radiation.

contaminated sites, broken down by area types—industrial sites, protected sanitary zones, and observation zones.

REGULATORY AND LEGAL BASE

The regulatory and legal base that has taken shape during the past decade has facilitated the safe operation of industry enterprises (see Figure 23-3). However, experience of its practical application also revealed problems associated primarily with the insufficient adaptation of the regulations to socioeconomic realities and tasks involved with promoting the country's sustainable development. One such problem was that Russian radiation hygiene regulations lacked special criteria for categorizing radiation and chemical contamination sites at nuclear industry enterprises. The initiation of special environmental programs and passage of Russian government resolutions served as an incentive for the development of such criteria.[7]

RULES FOR CATEGORIZING ROSATOM ENTERPRISES SITES SUBJECTED TO CHEMICAL AND RADIOACTIVE CONTAMINATION[8]

The rules for site categorization (hereafter referred to as "the rules") establish requirements for areas subjected to radioactive or chemical contamination or both as a result of the activities of Rosatom enterprises. They also set criteria that serve as the basis of decisions on the rehabilitation of sites; the conduct of ameliorative, site clearance, and soil treatment efforts on them; the establishment of protected zones; and the preservation of facilities located on such sites.

The rules are intended to define degrees of chemical and radioactive contam-

[7]Federal Law on Special Environmental Programs for Rehabilitating Radiation-Contaminated Sites, No. 92-FL, July 10, 2001. Federal Law on the Transfer of Land or Parcels of Land from One Category to Another, No. 172-FL, December 21, 2004. Resolution of the Russian Government on the Use of Land Subjected to Radioactive and Chemical Contamination, the Conduct of Ameliorative, Site Clearance, and Soil Treatment Efforts, and the Establishment of Protected Zones, No. 112, February 27, 2004.

[8]The rules were developed by the following coauthors: S. N. Brykin, I. S. Serebryakov, N. S. Roznova, M. V. Bufetova, S. A. Yakushev (Federal State Unitary Enterprise—All-Russian Scientific Research Institute of Chemical Technology); I. I. Linge, S. V. Kazakov, A. I. Ilyushkin, I. A. Osipyants, L. M. Vorobyova (Russian Academy of Sciences Nuclear Safety Institute); V. P. Tishkov, A. V. Stepanov, Yu. A. Panteleev, V. M. Gavrilov, S. K. Vasiliev, Ye. L. Lebedev (Federal State Unitary Enterprise—V. G. Khlopin Radium Institute Research and Production Association); I. I. Kryshev, A. I. Kryshev, T. G. Sazykina, K. D. Badagyan, M. N. Katkova, L. V. Makarova (Typhoon Research and Production Association—State Institution); L. A. Ilyin, N. K. Shandala, O. A. Kochetkov, M. N. Savkin, N. Ya. Novikova, S. G. Monastyrskaya, N. P. Sayapin, A. V. Titov, V. A. Seryogin (State Science Center—Institute of Biophysics); I. P. Korenkov (Radon Moscow Research and Production Association); and G. S. Perminova (Federal Monitoring Service for the Protection of Consumers' Rights and Human Welfare [Rospotrebnadzor]).

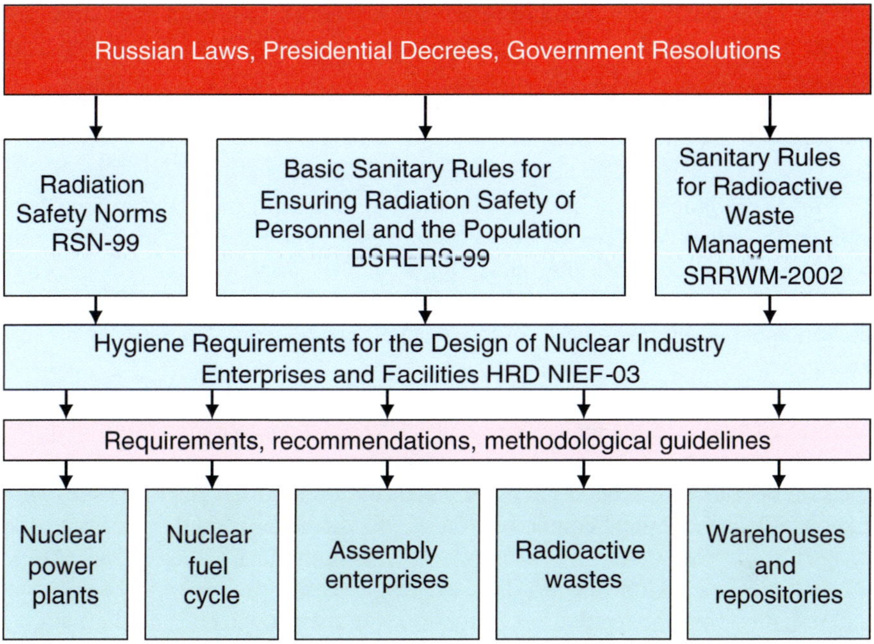

FIGURE 23-3 Regulatory-methodological base for ensuring radiation safety for Rosatom enterprise personnel and the population within the framework of sanitary legislation.

ination of sites and to determine procedures for their use, as follows, depending on the nature and level of contamination and on indicators of unfavorable effects on human health or the environment:

- Transfer of land into conservation preserve status
- Use for a specific purpose under special conditions
- Use for a specific purpose without any special conditions

The rules were developed for use by all legal entities and individuals owning contaminated territories, officials of the organizations operating radiation facilities, and organizations conducting environmental assessments and land rehabilitation. These rules apply to land and bodies of water located on the following territories:

- Sites contaminated as a result of unplanned situations and accidents at radiation facilities
- Areas freed up as a result of the reduction in size of protected sanitary zones around radiation facilities

- Previously occupied protected sanitary zones, industrial sites, and protected zones around radiation facilities that have now been removed from operation
- Protected sanitary zones, industrial sites, and protected zones of operating radiation facilities

In accordance with Article 55 of the Federal Law on the Sanitary-Epidemiological Welfare of the Population, responsibility for complying with these rules lies with the management of the various Rosatom enterprises (organizations). Compliance with the rules is monitored by federal executive branch agencies empowered to exercise state management of the use of nuclear energy.

The primary objective of land categorization is to identify, evaluate, and analyze current hazards and threats caused by existing contamination, with the aim of developing, introducing, and improving preventive measures to ensure the safe operation of related facilities. The criterion used to categorize sites with radioactive and chemical contamination of soil and water is the level of potential danger to the public and the environment. The numerical values of the criteria were developed taking into account existing Russian sanitary-epidemiological and radiation hygiene regulations. Attention was also paid to current international experience, particularly approaches to ensuring radiation safety laid out in the new recommendations of the International Commission on Radiological Protection (ICRP).[9]

The draft of the ICRP fundamental recommendations, which is being prepared to replace the previous fundamental recommendations laid out in 90 ICRP publications, develops a new approach to ensuring radiation safety in various exposure situations. The primary special feature of the draft recommendations lies in their departure from differentiating practical activities and intervention. Instead, the ICRP has proposed three exposure situations: planned, accidental, and existing. The category of existing exposure refers to unregulated ionizing radiation sources already existing when a decision is made regarding the need to monitor or control them. This pertains to natural sources such as radon, as well as previous nuclear activities considered in this work. The second feature of the new ICRP recommendations is their call for the optimization of radiation protection under boundary dose conditions in each case of exposure. Previously, boundary doses were used only as a control to monitor sources within the framework of the given practice. Under the new recommendations, the concepts of levels of intervention and levels of action no longer exist; they have been replaced by

---

[9]See Draft Recommendations of the International Commission on Radiological Protection, May 3, 2007, available online at *www.icrp.org*.

Assessing dose of the representative person for the purpose of radiation protection of the public and the optimization of radiological protection. ICRP Publication 101. Annals of the ICRP 36(3).

boundary doses, which are defined as the upper limits for the optimization of protection.[10]

CRITERIA FOR ASSESSING CHEMICAL CONTAMINATION OF SOIL AND BODIES OF WATER

Criteria for assessing soil contamination are set by the existing hygiene and environmental regulations, including maximum allowable concentration (MAC), reference allowable concentration (RAC), and reference allowable action level. MAC calculations include specific and integrated soil quality indicators. Soils are categorized according to their level of contamination: clean, allowable, moderately hazardous, hazardous, and extremely hazardous.[11] The degree of soil contamination is characterized by the total contamination indicator and the actual chemical substance content in the soil.

The total indicator for chemical contamination of soil makes it possible to determine the degree of chemical contamination of soils in sites being studied for the presence of harmful substances of various hazard classes. The concentration coefficient for a chemical substance is defined as the ratio of actual chemical substance content in the soil to the background value and, for contaminants of anthropogenic origin, as the quotient of the division of its mass share by the corresponding MAC. Regional soil indicators should be used as background values for the concentrations of chemical substances. The concept of regional background content of chemical substances entails their content in soils at sites not experiencing any technogenic load.

Indicators of the level of soil contamination are established depending on their further functional significance—that is, the purposes for which the rehabilitated land will be used. Classification of soils by their degree of contamination is presented in Table 23-1.[12]

Existing hygiene and environmental norms are used as criteria for evaluating the contamination of bodies of water, including the MAC and RAC for chemical substances in bodies of water, which have been approved according to established procedures.[13] For bodies of water used for communal drinking water and recreational purposes, the MAC for chemical agents in the water must not have a direct or indirect effect on the human body over an entire lifetime and must not impact the health of future generations. It must also not be detrimental to hygienic conditions for water use. For bodies of water used for commercial fishery purposes, the regulated standard is the MAC for the harmful substance in the water, which must not have a harmful effect on the fish population, primarily fish being

---

[10] Op. cit.

[11] Sanitary rules and norms 2.1.7.1287-03. Sanitary-epidemiological requirements for soil quality.

[12] Op. cit.

[13] Hygiene Norms 2.1.5.1315-03. Maximum allowable concentrations for chemical substances in bodies of water.

TABLE 23-1 Assessment of Levels of Chemical Contamination of Soils

		Content in Soil (mg/kg)					
		Hazard Class I		Hazard Class II		Hazard Class III	
Contamination Category	Total Indicator	Organic Compounds	Inorganic Compounds	Organic Compounds	Inorganic Compounds	Organic Compounds	Inorganic Compounds
Clean	—	From background to MAC	From background to MAC	From background to MAC	From background to MAC	From background to MAC	From background to MAC
Allowable	<16	1-2 MAC	From 2 backgrounds to MAC	1-2 MAC	From 2 backgrounds to MAC	1-2 MAC	From 2 backgrounds to MAC
Moderately hazardous	16-32	1-2 MAC	From 2 backgrounds to MAC	1-2 MAC	From 2 backgrounds to MAC	2-5 MAC	From 2 backgrounds to MAC
Hazardous	32-128	2-5 MAC	From MAC to Cmax	2-5 MAC	From MAC to Cmax	>5 MAC	> Cmax
Extremely hazardous	>128	>5 MAC	> Cmax	>5 MAC	> Cmax	>5 MAC	> Cmax

NOTE: Cmax indicates the maximum value for the allowable level of element content based on one of four hazard indicators.

commercially raised. In cases where a body of water is entirely or partially used simultaneously by various types of water users, the strictest quality regulations for surface waters must be taken into account, namely the norms for bodies of waters used for commercial fisheries.

For assessing the level of contamination in bodies of water, a special combined contamination index is used, taking into account not only actual concentrations of ingredients in the water but also multiple rate indicators and frequencies with which levels exceed the MAC. Water quality is classified according to the values achieved in the combined contamination index (see Table 23-2).

TABLE 23-2 Characteristics of Integrated Water Quality Assessment

Combined Contamination Index	Water Quality Class	Quality Assessment
≤0.2	I	Very clean
0.2-1.0	II	Clean
1.0-2.0	III	Moderately contaminated
2.0-4.0	IV	Contaminated
4.0-6.0	V	Dirty
6.0-10.0	VI	Very dirty
>10	VII	Extremely dirty

CRITERIA FOR ASSESSING RADIOACTIVE CONTAMINATION OF SITES

Radiation-contaminated sites are categorized according to mandatory compliance with existing legislative and regulatory-methodological documents.[14] Criteria for evaluating radiation-contaminated sites affected by radionuclides of technogenic origin are found in existing regulations on ensuring radiation safety for humans in these areas. Technogenic radionuclides include those that are artificially created as well as those of natural origin.

To determine radionuclide activity in the soil and water at sites subjected to radioactive contamination, studies are conducted and mandatory monitoring of the radiation situation is established. Status assessments are made of the territories around active radiation facilities after determining annual effective exposure

[14]Ministry of Health of the Russian Federation. 1999. Radiation safety norms RSN-99. Hygiene norms 2.6.1.758-99. 116 pp.

Ministry of Health of the Russian Federation. 2000. Basic Sanitary Rules for Ensuring Radiation Safety of Personnel and the Population BSRERS-99: Section 2.6.1 Ionizing radiation and radiation safety. Sanitary rules 2.6.1.799-99. 98 pp.

Sanitary Rules for Radioactive Waste Management SRRWM-2002. Sanitary rules 2.6.6.1168-02.

Federal State Unitary Enterprise—InterSEN. 2002. Hygiene safety requirements and nutritional values of food products: sanitary-epidemiological rules and regulations. Sanitary Rules and Regulations 2.3.2.1078-01. 168 pp.

doses to the public resulting from all technogenic sources and calculating the contribution of the contaminated sites in question to this dose.[15]

The exposure dose must be calculated in accordance with methodological guidelines established by the Main State Sanitary Physician of the Russian Federation (or the deputy) or by the Main State Sanitary Physician for the various organizations and jurisdictions served by the Russian Federal Medical-Biological Agency.

The assessment is made taking into account the goals for the use of the sites subjected to radioactive contamination and the means by which they were irradiated. Depending on the effective exposure dose to the public established for the average individual, sites may be categorized as follows:

- <10 μSv per year: A negligibly small level of risk, at which radiation sources do not fall under the scope of the Radiation Safety Norms (RSN-99), as they have no significant impact on public health.
- 10 μSv per year–0.2 mSv per year: In international and domestic practice, the level of 0.2 mSv per year is recommended as the level below which radiation sources should be exempt from regular monitoring.
- 0.2-1 mSv per year: The range of ongoing optimization, with the scope and nature of actions that are determined, taking into account the long-term forces for radiation impacts on the public.
- 1-10 mSv per year: Studies are conducted to clarify the radiation situation and select the optimal means of land use based on the principle of scientific evidence and optimization.
- >10 mSv per year: A decision is made on rehabilitating contaminated land, including within the bounds of protected sanitary zones and industrial sites.

In drawing up the scientific foundations for the categorization criteria, six scenarios for possible site use were considered along with their corresponding means of human exposure:

1. Constant human occupation of the contaminated site and full agricultural use

[15]Shandala, N. K., O. A. Kochetkov, and M. N. Savkin. 2004. Rehabilitation criteria for territories contaminated by man-made radionuclides. Proceedings of the Eleventh Congress of the International Radiation Protection Association, May 23-28, 2004, Madrid, Spain.

Ilyin, L. A., N. K. Shandala, and M. N. Savkin. 2004. The place and role of radiation hygiene monitoring in the sociohygienic monitoring system. Hygiene and sanitation 5:9-15.

Shandala, N. K., O. A. Kochetkov, M. N. Savkin, and N. Ya. Novikova. 2004. Radiation hygiene regulation in the problem of returning rehabilitated sites to economic use. Pp. 47-49 in Actual questions of radiation hygiene: A collection of papers from a scientific-practical conference, June 21-25, 2004. St. Petersburg: Russian Federation Ministry of Healthcare and Social Development.

2. Constant human occupation of the contaminated site and agricultural use for personal gardening only
3. Constant human occupation in an urban-type dwelling with no agricultural use
4. Operation of industrial enterprises and offices on the categorized site (temporary human presence at the contaminated areas)
5. Temporary human presence at the contaminated site for recreational purposes (estimated individual presence of 1 month per year on the site)
6. Agricultural use of the site (for crop growing and livestock grazing)

In connection with the substantial diversity of dose magnitudes resulting from various routes of exposure, it is impossible to calculate universal allowable levels of surface contamination or specific radionuclide activity in the soil.[16] Therefore, the rules set forth guideline magnitudes for radionuclide content in soil (Bq/m^2) at which an effective dose of 1 mSv per year is created, depending on the five scenarios for possible human activity (see Table 23-3). An additional condition that must be met for the sixth scenario is that the food items produced must not exceed certain values for specific radionuclide activity.[17]

The reference values cited in Table 23-3 may be used as soil-screening indicators in categorizing sites.

TABLE 23-3 Density of Soil Contamination in Bq/m^2 at Effective Dose to the Public of 1 mSv per Year for Various Site Use Scenarios

	Scenario Number				
Nuclide	1	2	3	4	5
Co-60	6.46E+04	6.63E+04	1.35E+05	4.56E+05	3.20E+05
Sr-90	2.15E+04	4.31E+04	2.61E+06	7.94E+09	1.41E+05
Cs-137	1.15E+05	1.22E+05	5.26E+05	1.81E+06	2.21E+05
Eu-154	1.32E+05	1.32E+05	2.65E+05	8.97E+05	6.32E+05
Ra-226	6.62E+04	9.47E+04	1.39E+05	8.82E+07	3.54E+07
Th-228	4.30E+05	4.36E+05	4.62E+05	8.45E+06	2.57E+07
Th-230	4.43E+05	4.50E+05	4.81E+05	2.84E+07	6.02E+07
Th-232	3.91E+05	3.97E+05	4.24E+05	1.59E+07	4.01E+07
U-233	7.21E+05	8.28E+05	1.42E+06	9.23E+07	1.87E+08
U-235	7.78E+05	8.93E+05	1.53E+06	1.07E+08	2.10E+08
U-238	8.43E+05	9.68E+05	1.66E+06	1.17E+08	2.29E+08
Pu-239	1.87E+04	1.87E+04	4.12E+05	7.94E+06	1.96E+04
Am-241	2.85E+05	2.87E+05	4.71E+05	9.45E+06	8.10E+05

[16]Levels of external exposure (due to radionuclides contained in various types of soils) and internal irradiation due to ingestion of radionuclides with locally produced food products, when the soil–food product transfer coefficients may vary by more than an order of magnitude depending on soil type.

[17]Federal State Unitary Enterprise—InterSEN. 2002. Hygiene safety requirements and nutritional values of food products: Sanitary-epidemiological rules and regulations. Sanitary Rules and Regulations 2.3.2.1078-01. 168 pp.

CONCLUSION

Based on the results obtained from categorizing radiation-contaminated areas, we may identify promising categories for land use and develop measures to rehabilitate land for particular purposes (agriculture, forestry, water supply, recreation, nature preservation, construction, and conservation). The transfer of industrial lands from one category to another is permitted, provided that there is an approved plan for land rehabilitation.

24

Areas of the Russian Federation Affected by Radiation Contamination Due to the Chernobyl Nuclear Power Plant Accident*

S. M. Vakulovsky, T. S. Borodina, A. A. Volokitin, V. M. Kim, G. I. Petrenko, E. G. Tertyshnik, A. D. Uvarov, and V. N. Yakhryushin,
Typhoon Research and Production Association—State Institution, Obninsk

As a result of the accident at the Chernobyl Nuclear Power Plant on April 26, 1986, 19 Russian Federation subjects (oblasts and autonomous republics) suffered cesium-137 contamination at levels of more than 1 Ci/km^2 (37 kBq/m^2). The total land area with this level of contamination in all the affected regions in 1986 was 65,050 km^2. Table 24-1 presents a 1998 forecast of expected changes in the contamination level over time due to radioactive decay.[1] As the table indicates, by 2006 the contaminated land area totaled 31,120 km^2—that is, it had been reduced in size by one-half during the 20 years since the accident.

All population centers located in areas with cesium-137 contamination levels of more than 1 Ci/km^2 (37 kBq/m^2) were studied to determine their level of radiation contamination.[2] The results of this research are presented in Table 24-2. From this table, it is clear that based on conditions in early 2006, there were 3,234 population centers in 15 Russian regions with cesium-137 contamination levels of more than 1 Ci/km^2 (37 kBq/m^2). For all these population centers, calcula-

*Translated from the Russian by Kelly Robbins.

[1]Izrael, Yu. A., et al. 1998. Long-term forecast of changes in the levels of Cs-137 radioactive contamination in areas of Russia after the Chernobyl Nuclear Power Plant accident. Meteorology and Hydrology 4:5-17.

[2]The term *population center* encompasses urbanized areas of various sizes ranging from large cities to towns, villages, and small rural settlements.

TABLE 24-1 Forecast of Expected Changes in Total Land Area in Russia with Varying Levels of Cesium-137 Radioactive Contamination Due to the Chernobyl Accident

	Area (in km^2) with Varying Levels of Local Cs-137 Contamination			
Year	>40 Ci/km^2	15-20 Ci/km^2	5-15 Ci/km^2	1-5 Ci/km^2
1986	580	2,070	5,780	56,260
1996	310	1,900	5,330	48,980
2006	40	1,280	3,540	26,260
2016		850	2,780	18,920
2026		625	2,700	15,040
2036		190	2,340	12,500
2046		100	1,500	10,930

tions were made to determine how long it would take for each to move to the next lowest contamination category as a result of radioactive decay. The results of these calculations are presented in Table 24-3, which indicates that there will be no population centers with contamination levels greater than 40 Ci/km^2 (1,480 kBq/m^2) by 2020, greater than 15 Ci/km^2 (555 kBq/m^2) by 2063, greater than 5 Ci/km^2 (185 kBq/m^2) by 2110, and greater than 1 Ci/Km^2 (37 kBq/m^2) by 2180.

Bryansk Oblast suffered the most contamination from the Chernobyl accident of any Russian region. Therefore, information was compiled on the contamination status of the near-earth atmosphere (up to approximately 5 km from the earth's surface) and bodies of water in this particular oblast. Figure 24-1 provides information on the change in the average annual cesium-137 activity by volume in the near-earth atmosphere over the city of Bryansk. From this figure, it is clear that the activity by volume decreased by an order of magnitude from 1987 to 1995. During the next 10 years, it changed insignificantly, fluctuating within the range of 1-3 × 10^{-6} Bq/m^3. This amount is seven orders of magnitude lower than the activity by volume allowable for the population, as established by the Radiation Safety Norms (RSN-99).[3] Levels of cesium-137 atmospheric fallout are cited for the population center of Krasnaya Gora (Figure 24-2), where the level of cesium-137 contamination is close to 5 Ci/km^2 (185 kBq/m^2), while near this town there is an area with substantially higher levels of soil contamination. Figure 24-2 shows that the level of atmospheric fallout declined by two orders of magnitude in the 5 years following 1986, changed insignificantly from 1991 through 2001, and subsequently fell to five times less by 2005. Figure 24-3 indicates that cesium-137 concentrations in the soil due to atmospheric fallout over 19 years totaled 11.5 kBq/m^2, which is 6.2 percent of initial contamination.

[3]Ministry of Health of the Russian Federation. 1999. Radiation security norms (RSN-99).

TABLE 24-2 Numbers of Population Centers in the Russian Federation by Level of Cesium-137 Contamination on an Oblast or Autonomous Republic Basis (Conditions as of January 2006)

Oblast or Autonomous Republic	Total Population Centers	Total Samples	<1 Ci/km²		1-5 Ci/km²		5-15 Ci/km²		15-40 Ci/km²		>40 Ci/km²	
			Population Centers	Sample	Population Centers	Sample	Population Centers	Sample	Population Centers	Sample	Population Centers	Sample
Bashkortostan	16	65	16	65								
Belgorod	550	3,597	419	3,007	131	590						
Bryansk	2,007	25,419	1,313	11,589	428	4,676	212	6,764	51	2.301	3	747
Volgograd	5	24	4	17	1	5						
Voronezh	1,208	9,636	1,136	8,901	72	735						
Kaluga	591	5,952	325	2,250	242	3,220	24	400				
Kursk	1,111	6,719	984	5,616	127	1,103						
Leningrad	159	1,732	122	1,335	37	430						
Lipetsk	214	1,607	170	1,281	44	375						
Mary El	25	72	25	72								
Mordovia	390	1,459	373	1,358	17	98						
Moscow	9	49	9	49								
Nizhny Novgorod	148	742	148	742								
Novgorod	85	492	85	492								
Oryol	1,596	11,850	1,001	7,201	594	4,640	1	9				
Penza	202	1,483	180	1,335	22	119						
Rostov	2	10	2	10								
Ryazan	579	7,365	405	5,871	174	2,493						
Saratov	11	41	11	41								
Smolensk	89	513	89	513								
Tambov	123	961	121	935	2	26						
Tula	2,391	19,005	1,340	7,943	995	9,855	56	1,207				
Ulyanovsk	131	559	130	552	1	4						
Chuvashia	34	89	34	33								
TOTAL	11,676	105,904	8,442	61,208	2,887	28,369	293	8,380	51	2,301	3	747

TABLE 24-3 Year When Last Population Center in Each Area Listed in Table 24-2 Will Shift from One Cesium-137 Contamination Category to the Next Lower

Oblast or Autonomous Republic	1-5 Ci/km^2	5-15 Ci/km^2	15-40 Ci/km^2	>40 Ci/km^2
Belgorod	2048			
Bryansk	2180	2110	2063	2020
Volgograd	2020			
Voronezh	2044			
Kaluga	2100	2030		
Kursk	2050			
Leningrad	2044			
Lipetsk	2039			
Mordovia	2040			
Oryol	2077	2007		
Penza	2050			
Ryazan	2054			
Tambov	2007			
Tula	2109	2039		
Ulyanovsk	2011			
TOTAL	2180	2110	2063	2020

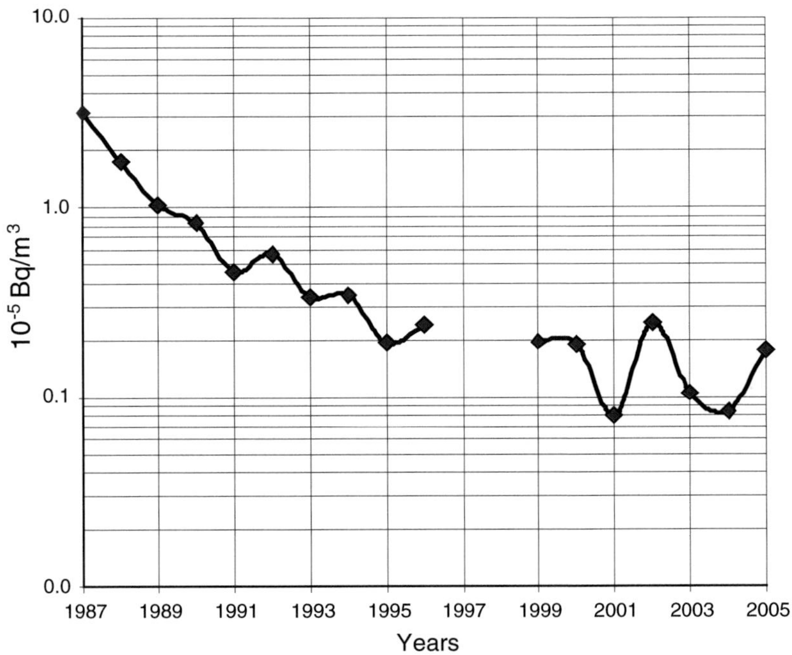

FIGURE 24-1 Changes in average annual cesium-137 activity by volume in the air over the city of Bryansk since the Chernobyl accident.

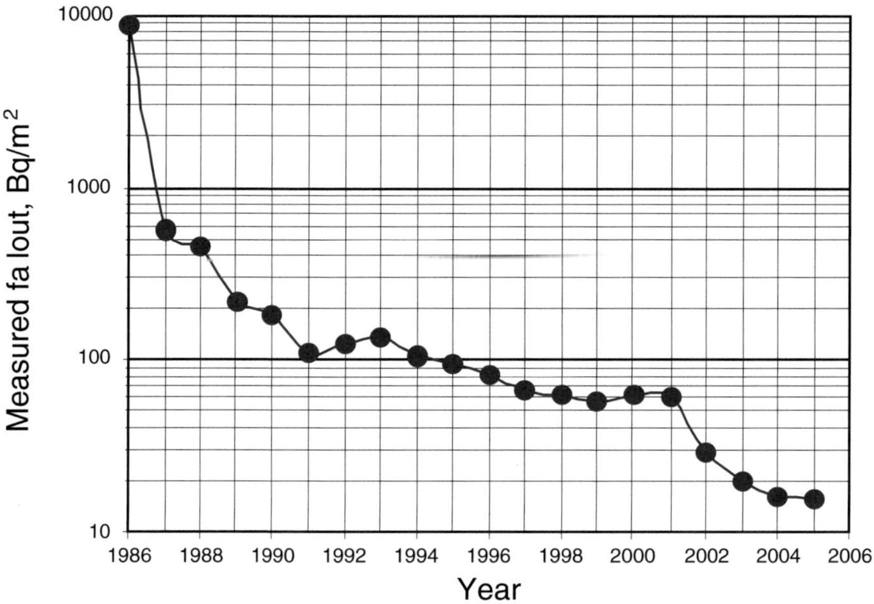

FIGURE 24-2 Annual cesium-137 fallout at Krasnaya Gora settlement.

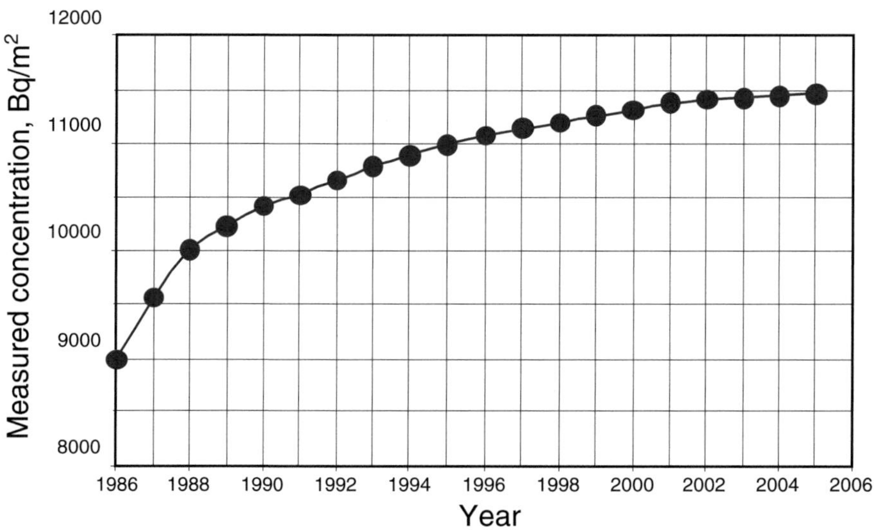

FIGURE 24-3 Cesium-137 concentrations in the soil in Krasnaya Gora settlement.

TABLE 24-4 Concentrations of Cesium-137, Strontium-90, and Tritium in Surface Water and Well Water, 1998-1999

	Concentration, Bq/L		
Body of Water	Cs-137	Sr-90	Tritium
Lake Svyatoye na Besedi	11.1-11.4	0.40-0.42	1.6-2.1
Lake Kozhanovskoye	5.1-7.1	1.1-1.5	1.4-3.0
Outflow canal from Lake Kozhanovskoye	4.4-5.9		
Nagorny Canal	0.5-1.4		
Karyer Reservoir	0.30-0.34	0.57-0.86	2.1-3.1
Ponds in rural settlements Yalovka and Nikolaevka	0.6-5.4	0.37-0.91	1.8-2.8
Other small bodies of water	0.07-3.54	0.17-0.71	1.2-4.2
Rivers Iput, Besed, Snov, Korna, Vaga, Tsata	0.01-0.23	0.021-0.095	1.9-2.7
Wells in rural settlement Nikolaevka	0.05-0.4		
Well in rural settlement Zaborye	0.08		
Well in center of rural settlement Kozhany	0.03		1.5
Borehole in urbanized settlement Mirny, school, water pipeline			0.2
RSN-99 levels of intervention, Bq/L	11	5	7,700

The levels of radioactive contamination in bodies of water in Bryansk Oblast were also studied in detail.[4] Some results of this research are presented in Table 24-4 and Figure 24-4. The table and figure show that levels of contamination in the water in rivers flowing through the contaminated areas of Bryansk Oblast and in wells located in the most contaminated population centers (Kozhany, Nikolaevka, and Zaborye) are two to three orders of magnitude below the level of intervention established by RSN-99. However, contamination levels in water in closed reservoirs located near those population centers (Lake Kozhanovskoye and Lake Svyatoye na Besedi) are close to or exceed the level of intervention. High levels of cesium-137 contamination in fish have also been identified in these lakes. Table 24-5 presents data on contamination levels in fish from these lakes. As shown in the table, the contamination levels in fish exceeded the maximum allowable level by one to two orders of magnitude.

[4]Vakulovsky, S. M., et al. 2000. Radioecological monitoring of the environment in Bryansk Oblast in 1998-1999. Pp. 19-23 in Proceedings of the International Conference on Radioactivity Associated with Nuclear Explosions and Accidents, April 24-26, 2000, Moscow, vol. 2.

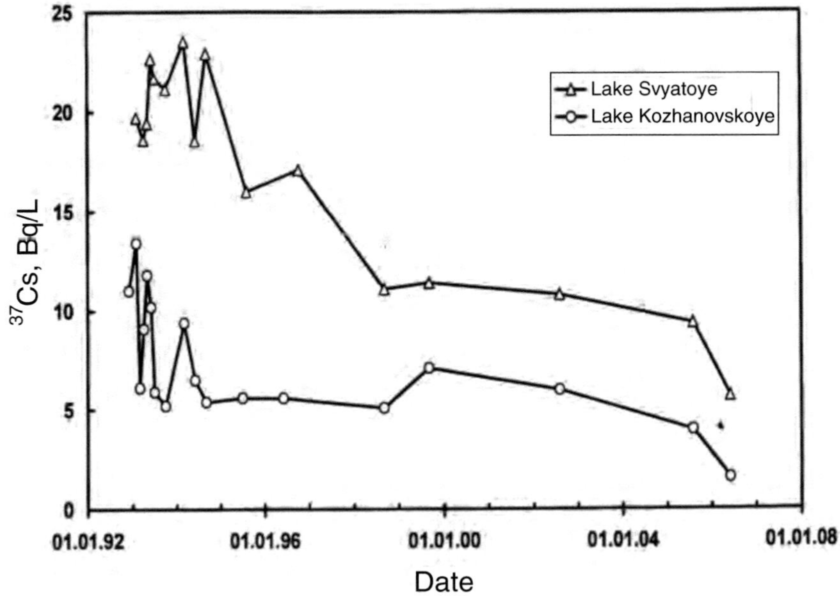

FIGURE 24-4 Cesium-137 concentration in lakes Svyatoye and Kozhanovskoye.

TABLE 24-5 Specific Cesium-137 Activity Levels in Fish, 1998-1999

Fish	Length, cm	Body of Water	Specific Activity, kBq/kg dry weight	Cs-137 Bioconcentration Factor in Fish, L/kg
Perch	5	Zalomenye	0.6	1,600
Carp	10	Yalovka	0.3	400
Carp	21-23	Vereshchaki	0.3	400
Carp	21-26	Karyer	0.1-0.3	300-900
Carp	25-30	Kozhanovskoye	5.2-9.5	1,100-2,100
Carp	20-36	Kozhanovskoye	6.1-10.0	1,300-2,200
Ruffe	10	Karyer	0.13	400
Northern pike	66	Karyer	0.7	2,200
Northern pike	17-43	Kozhanovskoye	5.5-12.4	1,200-2,700
Roach	10-15	Kozhanovskoye	2-5.8	400-1,300
Roach	22	Kozhanovskoye	6	1,300

25

The Experience of the Joint Environmental-Technological Scientific Research Center for Radioactive Waste Decontamination and Environmental Protection (MosNPO Radon) in Eliminating Radiation-Hazard Facilities and Rehabilitating Contaminated Sites*

V. G. Safronov, V. A. Salikov, Yu. A. Pronin, and S. V. Mikheikin

The following laws and regulations govern deactivation of contaminated areas in Russia:

- Federal Law 3-FL On Radiation Safety for the Population, January 9, 1996
- Federal Law 92-FL On Special Environmental Programs for the Rehabilitation of Radiation-Contaminated Parcels of Land, July 10, 2001
- Resolution on the Development of Special Environmental Programs for the Rehabilitation of Radiation-Contaminated Parcels of Land, approved by Resolution No. 421 of the Government of the Russian Federation, June 14, 2002
- Principles of State Policy on Ensuring Nuclear and Radiation Safety in the Russian Federation through 2010 and Beyond, approved by the President of Russia on December 4, 2003 (Pr-2196)
- Rules for the Use of Land Subjected to Radioactive and Chemical Contamination; the Conduct of Ameliorative, Site Clearance, and Soil Treatment Efforts; the Establishment of Protected Zones; and the Preservation of Dwellings and Production, Social, Cultural, and Public Service Facilities Located Thereon;

*Translated from the Russian by Kelly Robbins.

approved by Resolution No. 112 of the Government of the Russian Federation, February 27, 2004

VOLGINSKY FIELD (VLADIMIR OBLAST)

Volginsky Field is located on lands of the scientific testing base for experimental animals of the All-Russian Scientific Research Institute of Agricultural Radiology and Agroecology, not far from the settlement of Volginsky, Petushki Region, Vladimir Oblast.

From 1972 through 1982, the institute added cesium-137, strontium-90, thorium-232, and uranium-238 radionuclides to the soil of an experimental plot. In addition, the field also featured three buried vessels 1.5 m in diameter that were partially filled with radioactive waste. There were also two piles of soil measuring 10 m^3 each that were contaminated with cesium-137 and strontium-90. The maximum depth to which the radionuclides had penetrated was 0.5 m.

A detailed radiation study conducted by the Joint Environmental-Technological Scientific Research Center for Radioactive Waste Decontamination and Environmental Protection (Radon) in the summer of 2003 indicated the following:

- The area affected by radioactive contamination hot spots (experimental plots, vessels for solid and liquid radioactive waste, and adjacent plots and soil piles) totaled 2,200 m^2.
- The maximum depth of the contaminated soil layer was 0.3 to 0.4 m.
- The gamma-irradiation exposure dose intensity at the contaminated sites ranged from 15 to 2,100 mR per hour at a distance of 0.1 m from the soil surface and from 60 to 1,250 mR per hour at 1 m from the surface.
- The volume of radioactive wastes in the form of contaminated soil totaled 670 to 1,500 m^3.
- The average volume of radioactive waste in the form of construction debris (concrete vessels for the storage of solid radioactive waste) and metal items (liquid radioactive waste storage vessel) was 10 to 20 m^3.

In early 2007, more than 450 m^3 of radioactive waste was removed from this site. Work to decontaminate the site is ongoing.

SOLNECHNOYE LAKE (MOSCOW OBLAST)

The radioactive contamination site near Solnechnoye Lake was discovered in June 1985. It is located 50 m south of the lake near the city of Ramenskoye in Moscow Oblast and occupies an area of about 1.2 ha (150 × 80 m). The primary radionuclide contaminant is radium-226. Responsibility for the radioactive contamination has not yet been established.

In 1994 a radiation study of a solid-refuse dump site was conducted, involving the following operations:

- A gamma-radiation map indicating measurements according to a 2×2 m grid scheme
- Borehole gamma mapping
- Well drilling
- Spectrometric analysis of soil and water samples

Several radioactive waste burial sites were discovered during this radiation study. The gamma map of the distribution of exposure dose intensity at the soil surface (that is, at a distance of 0.1 m above the surface) indicated that radioactive contamination at the dump site was of a dispersed nature and not localized or grouped.

The gamma-irradiation exposure dose intensity at the surface of the radioactive contamination hot spots varied from 40 to 2,000 mR per hour. At the epicenter of one of the radioactive anomalies, the gamma-irradiation dose intensity at a depth of 0.7 m was more than 3,000 mR per hour. Radioactive waste was found buried mainly at a depth ranging from 0.5 to 4 m, with the maximum burial depth of up to 6 m.

Specific radium-226 activity varied from 3×10^3 to 5×10^5 Bq/kg. In water samples taken from Solnechnoye Lake and silt deposits from its shoreline, the radium-226 content did not exceed background values. The total amount of contaminated soil was estimated at 2,600 m^3.

In accordance with the Hygiene Requirements for Limitation of Public Exposure Due to Natural Sources of Ionizing Radiation (Sanitary Rule 2.6.1.1292-2003), a system was instituted for classifying industrial wastes according to the effective specific activity of natural radionuclides (see Table 25-1), and radiation hygiene requirements for their handling were established.

The technology for deactivation of the dump site included five stages:

1. Detailed radiation study of the site aimed at clarifying the location of surface radioactive contamination hot spots and evaluating the necessary scope of decontamination work; development of a plan for carrying out such work

TABLE 25-1 Categories of Industrial Waste Containing Natural Radionuclides

Waste Category	Effective Specific Activity (A_{eff}) of Natural Radionuclides, kBq/kg	Gamma-Irradiation Dose Intensity from Natural Radionuclides in Waste, mR/hr
Category I	$A_{eff} \leq 1.5$	$R \leq 70$
Category II	$1.5 < A_{eff} \leq 10.0$	$70 < R \leq 450$
Category III	$A_{eff} > 10.0$	$R > 450$

SOURCE: Hygiene Requirements for Limitation of Public Exposure Due to Natural Sources of Ionizing Radiation (Sanitary Rule 2.6.1.1292-2003)

2. Preparatory work
3. Decontamination work
4. Reclamation of decontaminated site and preparation of certificate of completion of decontamination work
5. Radiation ecology study of the dump site after elimination of surface radioactive contamination hot spots and site reclamation

During decontamination of this site, localized radioactive hot spots were discovered and subsequently eliminated to the point where remaining gamma-irradiation equivalent dose intensity values exceeded background levels by no more than 0.2 μSv per hour at a distance of 0.1 m from the reclaimed surface (rule 5.3.2 RSN-99).

From 2002 through 2006, Radon personnel removed about 160 m^3 of contaminated soil and sent it to long-term storage. Plans call for the removal of about 50 m^3 of additional radioactive waste in 2007.

BANK OF THE MOSCOW RIVER

The right bank of the segment of the Moscow River from Kolomenskoye Embankment to the railway bridge across the river near the Moskvorechye platform is one of the most unfavorable sites in the city of Moscow from a radioecological standpoint. This area is part of the Kolomenskoye State Museum Preserve and the Moskovorechye-Saburovo municipal district of the Southern Administrative District of the city of Moscow. In the layout of the city, the territories of the following institutions and enterprises adjoin this area (listed from north to south along the Moscow River): the Oncology Science Center, the Moscow Engineering and Physics Institute, the All-Russian Scientific Research Institute of Chemical Technology, the All-Russian Scientific Research and Design Institute of Industrial Technology, and the Moscow Polymetals Plant State Enterprise. On the grounds of these enterprises are a research nuclear reactor and radiochemical laboratories.

During their long operating histories, the industrial facilities of most of the enterprises listed above and a portion of the adjacent area along the bank of the Moscow River were contaminated by radioactive substances. In particular, beginning in the 1950s and through the 1980s, part of the ravines along the bank became covered with spent radioactive ores and waste (tailings) from radiochemical production facilities.

In recent years, work on this problem has become regular and systematic. On the whole, it is possible to point to the following types of work currently being carried out on the riverbank and being directly linked to or accompanied by measures for radioecological study, radiation monitoring, and deactivation:

- Efforts to decontaminate sedimentation vats at the neutralization station of the Moscow Polymetals Plant and adjacent areas

- Efforts to conduct a radioecological study associated with processes of riverbank erosion that uncovered radiation-contaminated spots that had previously been covered with clean soil
- Efforts associated with reconstruction of facilities on the grounds of the Kolomenskoye State Museum Preserve
- Land reclamation work to shore up the bank of the Moscow River
- Land reclamation work at sections of the bank not previously studied

In addition, there is currently no detailed and complete picture of the distribution of radionuclides across the entire bank area or the depth of their penetration.

Since 1997 a total of 10,548 m^2 of radiation-contaminated land has been discovered and decontaminated, and more than 1,300 m^3 of radioactive waste has been removed and transferred for long-term storage.

As a result of the studies, it has been established that the main contaminants are radionuclides belonging to the radium (radium-226) and thorium (thorium-232) families. The composition and specific activity values for a significant part of the contaminated soil make it possible to classify such soil as low-level and in some cases medium-level radioactive waste, which in turn makes it possible to consider this contaminated area as an uncontrolled tailings repository of a radiochemical production facility. Based on preliminary assessments, the amount of radiation-contaminated industrial waste (soil) at this site could be up to 60,000 m^3 (about 100,000 metric tons).

The main radioecological and other characteristics of this site are as follows:

- Large and expansive territory
- Presence on the site of a world-renowned facility with state museum preserve status, which is also a place of pilgrimage by many people, including foreigners
- Presence on the southern part of the riverbank in the Moscow Polymetals Plant observation zone of large volumes of radiation-contaminated industrial waste with significant specific activity levels
- Insufficient study of the remainder of the riverbank
- Constant probability of the appearance of new surface hot spots and radioactive contamination sites
- Possibility of pollution of the waters of the Moscow River from erosion processes and anthropogenic activities

TAININSKOYE VILLAGE (MOSCOW OBLAST)

A radiation-contaminated site was discovered on September 21, 2004, during a radiation ecology monitoring study in the village of Taininskoye, Mytishchi Region, Moscow Oblast. The site is located in riverside meadows, some of which are used for gardens, on the outskirts of the village on the left bank of the river Yauza.

As a result of a detailed radiation study of the site, which totaled 18,500 m^2 in area, it was discovered that background gamma-irradiation dose intensity levels typical for this area were within the range of 6 to 10 mR per hour, with the average background level being 8 mR per hour. A total of 25 radiation-contaminated hot spots were found during the study, with gamma-irradiation dose intensity levels of 29 to 220,000 mR per hour. All hot spots are of a localized nature, with the area of each not exceeding 0.01 m^2. The depth of deposition of the contaminants ranged from 0.1 to 0.8 m. The primary radionuclide contaminant according to laboratory spectrometric analysis data was radium-226.

Given that some of the radiation-contaminated hot spots were characterized by high levels of gamma-irradiation dose intensity (120-220,000 mR per hour), Radon specialists quickly decontaminated 12 hot spots. During this process, they separated out 240 kg of radioactive waste and subsequently sent it away for long-term storage.

The main characteristics of this particular site are as follows:

- Very small size of radiation-contaminated hot spots, with most being only 0.01 m^2 in area
- Location of contaminated soil predominately on the surface
- Use of part of the contaminated site for private agricultural activities
- Presence of garden plots, with access to the sites being coordinated with the plot owners
- Inaccessibility of a significant part of the contamination hot spots to vehicles or equipment
- Presence of dump sites for household refuse and thick vegetation, hindering study and deactivation efforts
- Lack of homogeneity of the contamination and presence of small fragments (<2-3 mm) containing radium-226

In developing the deactivation plan, it was determined that deactivation would be deemed complete if the gamma-irradiation dose intensity in excess of the background level did not exceed 30 mR per hour. If the 30 mR per hour level was exceeded, another 10-cm layer of soil would be removed, and if the gamma-irradiation dose intensity at the bottom of the resulting new indentation was less than 60 mR per hour, the deactivation effort would cease. Otherwise, the decision on the need for further deactivation would be made in consultation with the Center for State Sanitary-Epidemiological Monitoring of the city of Mytishchi.

The proposed deactivation completion criteria make it possible to ensure that in a 50-cm layer of contaminated soil there will be no mass greater than 5 kg and a gamma-irradiation dose intensity less than 5 mR per hour above background after rehabilitation.

As a result of Radon's activities, practically all surface contamination points discovered in Moscow have been eliminated. Meanwhile, ongoing characteristic problems still require attention. The experience Radon has acquired also attests to the city having radioactive contamination sites that have been hidden by new construction, eroded soil from ravines, and other construction work. Ensuring public safety demands that such hot spots be discovered and eliminated.

The fundamental factors tangentially or directly leading to the formation of radioactive anomalies are as follows:

- Absence during the 1940s through the 1960s of a legislative and legal base and scientifically founded regulations for radiation safety when dealing with radioactive materials
- Use of industrial waste, slag, and other materials containing radioactive substances in construction
- Lack of a centralized service for collecting and disposing of radioactive waste before 1961
- Imperfection of the system for accounting and control of radioactive substances at city enterprises
- Uncontrolled reorganization of enterprises working with radioactive substances and the transfer of their territories, facilities, and equipment to organizations and private individuals not associated with the use of radioactive substances

All of these reasons led to a situation in which, over the course of decades, various industrial wastes containing radioactive substances were hauled away from enterprises and organizations and dumped in uncontrolled fashion in vacant lots, ravines, and forests on the outskirts of the city or taken to dumps that did not observe radiation safety norms.

From 1971 through 2004, the focused efforts initiated by Radon to discover radioactive anomalies made it possible to identify and decontaminate 1,415 radioactive contamination sites in the city of Moscow and 256 sites in Moscow Oblast.

The most representative examples of large sites decontaminated by Radon in the city of Moscow include the following:

- A city block (approximately 0.7 km^2) contaminated by slag dust hauled in to fill and grade the site (maximum exposure dose intensity up to 2 R per hour)
- A site near the Moscow Ring Road (20,000 m^2, maximum exposure dose intensity up to 0.72 R per hour)
- Contaminated sites on kindergarten playgrounds
- Sites in recreational areas in the Kolomenskoye Preserve and the Izmailovo and Kuzminki Forest Parks (maximum exposure dose intensity up to 30 R per hour)
- A site in Victory Park on Poklonnaya Hill and the Olympic Village

Deactivation efforts in the city of Moscow have been carried out on the grounds of enterprises, in apartments, on playgrounds, at construction sites, and in forest parks, cultural and recreational parks, nature preserves, museums, and other areas where large numbers of residents and guests of the city assemble for recreation. The items being decontaminated have included both individual objects (ampoules; pieces of metal, concrete, and slag; instruments and their components; household items; oil pipes; and so forth) and localized (from 1 to 10 m^2) and more expansive (>10 m^2) areas of radioactive contamination in buildings and on their grounds.

The past 20 years have seen a serious strengthening of the regulatory and legislative base for operations involving radioactive substances. Such changes have meant that radioactive waste previously buried at industrial dump sites must be removed, processed, and buried in accordance with current requirements. Since the time when a number of nuclear industry enterprises began operating in the 1940s, the city of Moscow has increased in size many times over, meaning that historical or unauthorized radioactive waste burial sites previously located outside the city limits are now located in densely populated areas. Furthermore, the city has no complete database on enterprises that previously worked with ionizing radiation sources but were closed or restructured as a result of economic policy changes during the past 20 years. Most of them changed ownership, some exist without any owners, and documentation on ionizing radiation sources has been lost.

26

Use of GIS Technology for Assessing Territories Contaminated with Radioactive Materials

A. N. Plate and A. V. Vesselovsky,
Russian Academy of Sciences Institute of Ore Deposits and Geology

The geographic information system Radiation Safety of Russia (RSR) is being used at the Russian Academy of Sciences (RAS) Institute of Ore Deposits and Geology (IGEM) in accordance with a state program. RSR is designed to facilitate the monitoring, analysis, and simulation of nuclear and radiation hazards in various regions of the country. Using this system, information may be exchanged with other data systems. The system maintains the data in digital form and provides insights on the radiation situation in Russia as a whole, in adjacent territories, and in specific regions, oblasts, and districts. Thus, it provides information of broad interest concerning radioecological monitoring.

The RSR's cartographic database, one of its central features, includes a set of heteroscale digital maps with indications of potentially hazardous objects. Data on objects in specific territories are presented in different layers (see Table 26-1).

The following types of digital maps show radioecological loads:

- Major radiation-hazard sites are shown on the digital topographic maps of Russia and adjacent countries (scale 1:5,000,000). These sites include enterprises involved in the nuclear fuel cycle (uranium deposits, chemical complexes, metallurgical and radiochemical plants, and nuclear power plants), nuclear scientific research and training centers, enterprises with nuclear research reactors, nuclear test sites and locations of underground peaceful nuclear explosions,

TABLE 26-1 Cartographic Database

Object	Number of Locations
Uranium deposits	40
Plants and complexes of the nuclear fuel sector	17
Facilities with nuclear reactors (power, research, other)	69
Base ports for nuclear-powered ships	15
Shipbuilding and ship repair facilities	14
Nuclear weapons enterprises	7
Peaceful nuclear explosions	62
Nuclear test sites	8
Radiation-related accidents	10
Chernobyl accident consequences	11
Kyshtym accident—East Urals Trace	4
Radioactive waste storage sites and regional repositories	51
Concentrations of fission products of radioactive elements from reactors	25
Major nuclides from uranium ore reprocessing wastes and refining concentrates	13

bases for nuclear-powered ships, nuclear weapons plants, and radioactive waste disposal sites on land and at sea (263 site descriptions).

- A digital map of the Barents and Karsk seas at a scale of 1:5,000,000 (G. G. Matishev, D. G. Matishev, and V. V. Nazimov, RAS Murmansk Marine Biological Institute) serves as an illustration of levels and main directions of radionuclide transfer in these bodies of water. The graphic database also includes information on cesium-137, cobalt-60, and plutonium-239/240 in bottom sediments, as well as cesium-137 in the water, lichens, soil, and fish.
- A digital map presents radioactive contamination from cesium-137 in the European part of Russia, the Urals region, Ukraine, Belarus, Moldova, the Baltic countries, and western Georgia (scale 1:5,000,000). The Institute of Global Climate and Ecology of the Russian Federal Service for Hydrometeorology and Environmental Monitoring (Rosgidromet) and RAS collected this data in 1993.
- Digital maps of the East Urals Radioactive Trace (EURT) (edited by V. N. Chukanov, Ekaterinburg) at scales of 1:1,000,000 and 1:200,000 illustrate strontium-90 distribution in Kamensk Region at the time of an accident and later. Also included is the first officially published EURT status map with data provided by the Mayak enterprise (scale 1:1,000,000). It illustrates reconstructed levels of initial strontium-90 contamination of Kamensk Region as well as current levels in the area (scale 1:200,000).

Table 26-2 presents information about nuclear fuel cycle operations.

The RSR is being updated as new operating conditions develop. The following activities are of interest: estimation of radionuclide migration direction and intensity depending on the lithological, geochemical, and hydrological situation;

TABLE 26-2 Radioactive Products Resulting from Nuclear Fuel Cycle Operations

Nuclear Fuel Cycle Operation	Radioactive Products	Accumulation Site	Radioactive Elements	Product Type	Product Quantity and Activity	Location
Mining	Uranium ore	Tailings pond	Ra-226, Rn			Uranium ore deposits
Reprocessing	U oxide of nuclear purity	Tailings pond	Ra-226, Rn			Mining-chemical complexes
	U hexafluoride	Steel vessels	HF (toxic at destruction)			Six plants in Angarsk, Tomsk, Yekaterinburg, and Krasnoyarsk
	Enrichment by U-235 isotope		U-235			
Production of fuel, fuel rods, and complete products	U hexafluoride, dioxide, tablets, TEEs, thermo-emitting assemblies		U hexafluoride		~1.6×10^6 m^3 ~9.3×10^4 Ci	Novosibirsk Chemical Concentrates Plant, Electrostal Mechanical Plant
Radiochemical reprocessing of nuclear materials	Enriched U, irradiated Np-237, weapons-grade Pu		U, Np-237, Pu	Liquid (high-, medium-, low-level), solid	From $n \times 10^6$ Ci up to $n \times 10^8$Ci from 0.1γ up to >1,000γ	Chelyabinsk-65 (Mayak), Siberian Chemical Complex (Tomsk), Krasnoyarsk-26
Spent fuel regeneration	Uranyl nitrate, U monoxide-oxide, Pu dioxides, Np-237		U, Pu, Np			RT-1 Plant (Mayak)
Production from residual solutions			Sr-90, Cs-137, Am, Tc and other	Liquid, solid, gaseous		RT-1 Plant (Mayak)
Secondary fuel production	Commodity U (alloy of hexa-anhydrite nitrate uranyl), Pu dioxide	Repository	U, dioxide Pu		30 metric tons of Pu	RT-1 Plant (Mayak)
Pu recycling for energy material ("quick neutrons")	U-Pu fuel				400 fuel rods fabricated	Mayak pilot plants

environmental changes at uranium deposits; and identification of geological conditions favorable for radioactive waste disposal.

The expanded system can handle the analysis and simulation of radiation-hazard situations. It allows for the accumulation of insights on radioactive contamination sources and can present information in the form of separate graphic (mapping) layers or combinations of layers. The system includes a mapping database featuring a set of digital topographic maps of various scales with indications of potentially hazardous facilities and contaminated areas. Facility and site characteristics are compiled in a graphic database, and insights may then be obtained with the help of interfaces with the appropriate layer and the object of interest on the screen. According to the digital topographic map of Russia and adjacent countries (scale 1:5,000,000), 263 sites present potential radiation hazards.

For example, radionuclide transfer in the waters of the Barents and Karsk seas are of concern. The overall activity of liquid radioactive waste dumped into the Barents Sea from 1960 to 1990 is about 5,000 Ci. From 1964 to 1990, about 11,000 containers of radioactive waste with a total activity of more than 175,000 Ci were deposited in the Novaya Zemlya Depression.

Nuclear accidents and disasters are another example of thematic coverage offered by the RSR. The Kyshtym disaster of September 29, 1957, led to the East Urals Trace, which extends more than 1,000 km with a width of 10 km. The total radioactivity discharge was about 20 million Ci. As a result of the Chernobyl accident, 50 million Ci of various nuclides and 50 million Ci of radioactive gases were discharged. About 1.5 million ha of Russian territory were contaminated by cesium-137 and strontium-90.

In connection with the database, reference material is provided for specific sites, such as characteristics of liquid, solid, and gaseous wastes; major radionuclides in the wastes from uranium ore processing and concentrate refining; and fission products generated as a result of reactor operations. Thus, the particular characteristics of technological cycles causing possible alterations of the ecological situation are taken into account.

REFERENCES

Bulatov, V. I. 1993. Two hundred nuclear test sites of the USSR: geography of radiation disasters and contaminations. Novosibirsk, 16 pp.

Chukanov, V. N., ed. 1996. East Urals Radioactive Trace. Ekaterinburg, 167 pp.

Egorov, N. N., V. F. Kudryavtsev, B. V. Nikipelov, et al. 1993. Regeneration and localization of nuclear fuel cycle radioactive wastes. Atomic Energy 74(4):307-312.

Matishov G. G., D. G. Matishov, V. V. Podobelov, and L. G. Pavlova. 1993. Radiation situation at Kola Peninsula, Novaya Zemlya, Zemlya Franz-Iosif and at Barentsevo Sea water area. Reports of the Academy of Science 330(4).

RAS Murmansk Marine Biology Institute. 1994. Map (one sheet). St. Petersburg.

Rosgidromet and RAS Institute of Global Climate and Ecology. 1993. Chernobyl Accident Consequences (Map). Moscow.

USSR National Report to 1992 United Nations Conference on Environment and Development: Project. 1991. Moscow, 193 pp.

Appendixes

Appendix A

Workshop Agenda

CLEANING UP SITES CONTAMINATED WITH RADIOACTIVE MATERIALS INTERNATIONAL WORKSHOP

Russian Academy of Sciences, Presidium, Moscow

June 4-6, 2007

June 4

Opening Session

Nikolay P. Laverov, Vice President, Russian Academy of Sciences (RAS)

Frank L. Parker, Vanderbilt University and Representative of the National Academies

Andrei B. Malyshev, Deputy Director, Federal Atomic Energy Agency (Rosatom)

Norbert Jousten, Executive Director, International Science and Technology Center (ISTC)

David N. McNelis, University of North Carolina and Representative of the Russell Family Foundation

Overview Session

Chair: Academician Boris F. Myasoyedov, RAS

A. M. Agapov, Rosatom: *The Environmental Policy of the Russian Federal Atomic Energy Agency (Rosatom) and Priority Objectives for its Implementation*

V. V. Shatalov, All-Russian Research Institute of Chemical Technology, Rosatom: *Environmental Protection Aspects of the Federal Targeted Program "Uranium of Russia"*

A. A. Sarkisov, RAS Nuclear Safety Institute: *Remediation of Radioactively Contaminated Sites as Part of the Strategic Master Plan for Nuclear Submarine Decommissioning*

S. N. Kalmikov, RAS V. I. Vernadsky, Institute of Geochemistry and Analytical Chemistry: *Radionuclide Speciation for Elaboration of Remediation Methods and Technologies*

V. I. Velichkin, RAS Institute of Geology of Ore Deposits, Petrography, Mineralogy, and Geochemistry: *Evaluation of Radiation Ecology Status Around Russian Nuclear and Radiation Enterprises Based on Landscape-Geochemical Research*

S. A. Dmitriev, Moscow Joint Environmental-Technological Scientific Research Center for Radioactive Waste Decontamination and Environmental Protection (MosNPO Radon): *Comprehensive Resolution of the Problem of Radioactive Waste Management and Rehabilitation of Contaminated Areas in the Moscow Region*

Case Studies

Chair: Vasily I. Velichkin, RAS

Case Study 1. V. P. Karamushka, All-Russian Research, Design, and Surveying Institute of Production Technology (VNIPIPT), Rosatom: *Lands Damaged as a Result of Uranium Ore Mining Operations in the Russian Federation*

Review of Report by U.S. Experts: J. H. Clarke and F. L. Parker

Discussion

Case Study 2. Ye. G. Drozhko, Mayak Production Association, Rosatom: *Experience in Rehabilitating Contaminated Land and Bodies of Water Around the Mayak Production Association*

Review of Report by U.S. Experts: B. Looney and F. L. Parker

Discussion

June 5

Case Studies (Continued)

Chair: Frank L. Parker

Case Study 3. V. G. Volkov, Russian Research Center—Kurchatov Institute: *Remediation of Contaminated Facilities at the Kurchatov Institute*

Review of Report by U.S. Expert: R. Gephart

Discussion

Case Study 4. Ye. N. Kamnev, VNIPIPT: *Industrial Nuclear Explosion Sites in the Russian Federation: Recovery and Institutional Monitoring Problems*

Review of Report by U.S. Expert: D. Bradley

Discussion

Case Studies (Continued)

Chair: Academician Ashot A. Sarkisov, RAS

Case Study 5. A. P. Vasiliev, International Center for Environmental Safety, Rosatom: *The Past, Present, and Future of the Facilities at Andreev Bay*

Review of Report by U.S. Expert: D. Rudolph

Discussion

Case Study 6. Yu. Ye. Gorlinsky, Russian Research Center—Kurchatov Institute: *Environmental Remediation of Spent Nuclear Fuel and Radioactive Waste Temporary Storage Facilities in Gremikha Village: Challenges and Proposed Solutions*

Review of Report by U.S. Expert: D. Rudolph

Discussion

June 6

ISTC Projects Related to Remediation of Radioactively Contaminated Lands

Chair: Waclaw Gudowski, ISTC

G. G. Kulikov, ISTC: *An Overview of ISTC Projects Concerning Problems of Radioactively Contaminated Land Remediation*

V. K. Popov, Russian Research Center—Kurchatov Institute: *Systems Studies of the Radiation Legacy and the Development of the Informational, Legal, and Regulatory Framework for Post-Rehabilitation Institutional Control, Oversight, and Management of Radiation-Hazard Facilities in the Russian Federation*

V. I. Torgashov, Scientific Research Institute of Physical and Chemical Problems, Belarus State University: *The Cleanup of Radioactively Contaminated Lands Based on a Nonradioactive Cellulose Technology*

A. V. Konoplev, Typhoon Research and Production Association, Russian Federal Service of Hydrometeorology and Environmental Monitoring (Rosgidromet): *The Development of Meliorant-Sorbent Composition and Production Technology for Remediation of Soils Contaminated with Radionuclides and Forecasting of their Application Effectiveness*

V. A. Kamachev, Khlopin Radium Institute, Rosatom: *Decontamination of Soils in Supercritical and Liquid CO_2 and in Ozone-Friendly Freon HFC-134a*

V. V. Toropova, Aristotle University, Thessaloniki, Greece: *On the Effect of Fe(III) and Cr(III) Hydroxo-Complexes on the Process of Chemical Decontamination of Soils Polluted with Cesium and Strontium Radionuclides as a Result of the Chernobyl Accident*

General Problems of Remediation of Radioactively Contaminated Sites

Chair: Sergei M. Vakulovsky, Rosgidromet

S. M. Vakulovsky, *Typhoon Research and Production Association, Rosgidromet: Areas of the Russian Federation Affected by Radiation Contamination Due to the Chernobyl Nuclear Power Plant Accident*

V. G. Linnik, RAS V. I. Vernadsky Institute of Geochemistry and Analytical Chemistry: *The Landscape Approach to Remediation of Radioactively Contaminated Sites: Development of a Decision Support System for Remediation Management*

V. Novikov, Russian Research Center—Kurchatov Institute: *IIASA Studies of Radioactivity Transfer from Contaminated Sites: The Modeling of Atmospheric*

Transfer from the Russian Far East and Runoff Transfer from the Storage Site at the Kurchatov Institute

Ye. V. Kvasnikova, RAS and Rosgidromet Institute of Global Climate and Ecology: *An Estimate of the Consequences of Excavation Activities at Sites of Nuclear Explosions After 40 Years*

General Problems of Remediation of Radioactively Contaminated Sites (Continued)

Chair: Yevgeny N. Kamnev, Rosatom

N. K. Shandala, Institute of Biophysics, Federal Medical and Biological Agency: *Criteria for Categorizing Territories at Russian Federal Atomic Energy Agency Enterprises Experiencing Chemical and Radioactive Contamination*

S. V. Mikheikin, MosNPO Radon: *The Experience of the Joint Environmental-Technological Scientific Research Center for Radioactive Waste Decontamination and Environmental Protection (MosNPO Radon) in Eliminating Radiation-Hazard Facilities and Rehabilitating Contaminated Sites*

A. I. Rybalchenko, All-Russian Research, Design, and Surveying Institute of Industrial Technology, Rosatom: *Rehabilitation of Contaminated Groundwater Layers Near the Mayak Enterprise Using Deep Burial Technology*

A. M. Sobolev, Research Institute of Atomic Reactors, Rosatom: *The Choice of Criteria and Elaboration of Measures Aimed at Remediation of Contaminated Land on the Site of the All-Russian Research Institute of Biological Protection of Plants in Krasnodar*

Concluding Summaries

Nikolay P. Laverov, RAS

Frank L. Parker, Vanderbilt University

Appendix B

Titles of Additional Papers and Extended Abstracts Presented at the Workshop on Cleaning Up Sites Contaminated with Radioactive Material

National Academies and Russian Academy of Sciences
June 4-6, 2007, Moscow

The authors should be contacted directly to obtain copies of their presentations.

Brykin, S. N., Yu. Ye. Gorlinsky, V. A. Kutkov, O. A. Nikolsky, V. I. Pavlenko, Yu. V. Svintsev, and B. S. Stepennov (Russian Research Center—Kurchatov Institute). *Selection of the Final Condition and the Strategy of Ecological Rehabilitation of the Temporary Storage Site for Spent Nuclear Fuel and Radioactive Waste in Gremikha Village.* Contact: yury@quest.poly.kiae.su.

Konoplev, A. V., and L. N. Maskalchuk (Typhoon Research and Production Association; Joint Institute for Power and Nuclear Research—Sosny). *The Development of Technologies for Increasing Soil Sorption Capacity for Rehabilitation of Soils Contaminated with Radionuclides and Assessment of Their Efficiency.* Contact: konoplev@obninsk.com.

Kryshev, I. I., A. I. Nikitin, and I. A. Ryabtsev (Typhoon Research and Production Association; Institute of Ecology and Evolution). *Radioecological Monitoring of the Techa-Ob River System and Remediation Problems of Contaminated Reservoirs.* Contact: kryshev@typhoon.ru.

Kvasnikova, Ye. V., S. K. Gordeev, and Ye. M. Gorbunova (Institute of Global Climate and Ecology; MosNPO Radon; Institute of Geosphere Dynamics). *An*

Estimate of the Consequences of Excavation Activities at Sites of Nuclear Explosions after 40 Years. Contact: ranexa.kv@q.23.relcom.ru.

Kulikov, G. G. (International Science and Technology Center). *An Overview of ISTC Projects Concerning Problems of Radioactively Contaminated Land Remediation.* Contact: www.istc.ru.

Linnik, V. G., A. G. Volosov, V. N. Potapov, A. A. Saveliev, A. V. Sokolov, P. V. Sokolov, and V. V. Surkov (Vernadsky Institute of Analytical Chemistry and Geochemistry; Russian Research Center—Kurchatov Institute; Ecological Faculty of Kazan University; Geography Faculty of Moscow State University). *The Landscape Approach to Remediation of Radioactively Contaminated Sites: Development of a Decision Support System for Remediation Management.* Contact: linnik@geokhi.ru.

Sarkisov, A. A. (Institute for Nuclear Safety). *Remediation of Radioactively Contaminated Sites as Part of the Strategic Master Plan for Nuclear Submarine Decommissioning.* Contact: sarkisov@ibrae.ac.ru.

Shadrin, A. Yu., A. A. Murzin, V. A. Kamachev, and D. N. Shafikov (Khlopin Radium Institute). *Decontamination of Soils in Supercritical and Liquid CO_2 and in Ozone-Friendly Freon HFC-134a.* Contact: shadrin@atom.nw.ru.

Sobolev, A. M. (Scientific Research Institute of Atomic Reactors). *The Choice of Criteria and Elaboration of Measures Aimed at Remediation of Contaminated Land on the Site of the All-Russian Research Institute of Biological Protection of Plants in Krasnodar.* Contact: sam@vinf.ru.

Torgashov, V. I., F. N. Kaputsky, O. V. Zubets, E. V. Gert, H. P. Fink, and D. L. Jones (Physical Chemical Research Institute of Belarus State University; Frauenhofer Gesellschaft e. v. Institute für Angewandte Polymerforschung; School of Agriculture and Forest Science of University of Wales). *The Cleanup of Radioactively Contaminated Lands Based on a Non-Radioactive Cellulose Technology.* Contact: Cell@niks.by.

Toropova, V. V., Yu. P. Davydov, N. I. Voronic, D. Yu. Davydov, L. M. Zemskova, A. S. Titov, and Panagiotis Misaelides (Joint Institute for Power and Nuclear Research—Sosny; Department of Chemistry of Aristotle University, Thessaloniki). *On the Effect of Fe(III) and Cr(III) Hydroxo-Complexes on the Process of Chemical Decontamination of Soils Polluted with Cesium and Strontium Radionuclides as a Result of the Chernobyl Accident.* Contact: Ireplob06@mail.ru.

Zhemzhurov, M. L., V. V. Skurat, R. A. Shaternik, N. K. Vientyeva, and Ye. V. Likhavets (Joint Institute for Power and Nuclear Research—Sosny). *Removal of Radiation Danger on the Former Sites of Military Units within the Territory of the Republic of Belarus.* Contact: irep@sosny.bas-net.by.